THE MOON OBSERVER'S HANDBOOK

The Full Moon (Lick Observatory photograph)

The Moon observer's handbook

Fred W. Price

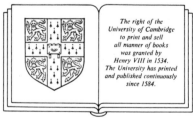

The right of the
University of Cambridge
to print and sell
all manner of books
was granted by
Henry VIII in 1534.
The University has printed
and published continuously
since 1584.

CAMBRIDGE UNIVERSITY PRESS

Cambridge
New York   New Rochelle   Melbourne   Sydney

Published by the Press Syndicate of the University of Cambridge
The Pitt Building, Trumpington Street, Cambridge CB2 1RP
32 East 57th Street, New York, NY 10022, USA
10 Stamford Road, Oakleigh, Melbourne 3166, Australia

First published 1988

Printed in Great Britain

British Library cataloguing in publication data

Price, Fred W. (Fred William), *1932–*
The moon observer's handbook.
1. Moon. Surface features. Observation –
Amateurs' manuals
I. Title
523.3

Library of Congress cataloguing in publication data
Price, Fred W. (Fred William), 1932-
The moon observer's handbook / by Fred W. Price.
    p.   cm.
Bibliography: p.
Includes index.
ISBN 0 521 33500 0
1. Moon—Amateurs' manuals. 2. Moon—Observers' manuals.
I. Title.
QB601.9.P75 1988
523.3–dc19

ISBN 0 521 33500 0

MU

To my mother
Corona Alexandra Price
and to the memory of my father
William George Price

CONTENTS

Contents

PLATES

FOREWORD

For a while amateur astronomers seemed to have lost interest in the Moon, following its exploration by space probes and astronauts, most likely because of an erroneous impression that we 'knew all there was to be known' about Earth's only natural satellite. This seems no longer to be the case, for various amateur groups are carrying out interesting and vigorous observing programs.

Dr Price, a most careful and self-critical observer, has drawn on his many years of lunar observation to produce a detailed guidebook to lunar observing, which fills an important gap and is a welcome addition to the literature. Anyone who wants to learn how to observe the Moon will surely profit from this work.

Ernst E. Both
Director and Curator of Astronomy,
Buffalo Museum of Science   New York   USA

PREFACE

This book has been written to help amateur observers of the Moon to observe effectively, to record their observations accurately and, where possible, to make contributions of scientific value. Several books have been written on observational astronomy containing chapters on lunar observation and the British Astronomical Association has produced its own *Guide* for the lunar observer. Many books are devoted to the detailed topography of the Moon, such as *The Moon* by H.P. Wilkins and P.A. Moore (Faber and Faber, 1955). However, there does not seem to be a comprehensive book-length manual devoted solely to lunar observation. The present work is an attempt to fill this gap.

Ever since the publication of the close-up photographs of the lunar surface obtained with the lunar probe and Orbiter space craft cameras and the preparation of accurate photographic lunar atlases, there has been a feeling in amateur astronomical circles that there is not much point in carrying on with Earth-based telescopic study of the Moon: it would seem hardly likely that data of scientific value are likely to result. However, I do not agree with this point of view for reasons detailed in the Introduction.

Lunar observation with Earth-based telescopes of the type found in amateur hands is by no means pointless or obsolete and this book is addressed to amateurs who study the Moon for their own pleasure as well as to observers who wish to contribute to knowledge of the Moon.

The Introduction suggests some answers to the question 'Why observe the Moon?' and the newcomer to lunar observation may thereby be helped to understand and clarify his own motivations and to decide whether he really wants to specialise in this area as well as demonstrating that Earth-based lunar observation is a worthwhile scientific pursuit. Following this, Chapter 1 gives the beginner a grasp of the relation of the Moon to the Earth, and the Moon's surface features, and the following chapter provides necessary 'background knowledge' of the Moon's motions in relation to the Earth and Sun. Chapter 3 describes the principle instrument of observation–

the optical telescope–and initiates the amateur into the essential elementary theory of the telescope so that he can use the instrument intelligently after deciding what sort of telescope to acquire. Chapter 4 is a guide to the main features of the Moon's surface as they successively come into view on consecutive evenings from new Moon to just after full Moon. Observational technique, getting to know one's way around the Moon and the best times of the day, month and year to observe are fundamental to serious lunar observation and these topics are dealt with in Chapter 5. At this point the amateur who wishes to do no more than to observe and record what he sees will probably be satisfied.

The last two chapters will interest serious amateurs who want to move on from the aesthetic satisfaction of lunar observation to more advanced scientific investigations. The chapter on 'Mysterious Happenings on the Moon' and the references elsewhere in the book to anomalous happenings and phenoma on the Moon are deliberately inserted because they are controversial and because a sense of mystery is one of the most powerful stimuli to investigation. Whatever light can be thrown on some of these lunar 'mysteries' by the skilled amateur will be a real contribution to knowledge. With this in mind the last chapter offers suggestions for serious research. Finally, the Appendices contain data likely to be useful to the practical lunar observer.

I have not compiled a comprehensive map of the Moon to illustrate this book because there are so many good ones in existence and I saw no point in duplicating what is already easily available. A list of Moon maps will be found in the Appendices. The usual small scale map of the Moon shows all types of formations and some of these maps are quite overcrowded with the names of surface features so that the effect is often to confuse the beginner. Instead of a single map, I have included a number of simple outline maps, each one devoted to one kind of lunar feature, e.g., craters are shown in one map, mountain ranges in another and so on. By so doing I hope to have overcome the crowded and confusing effect of one comprehensive map and have enabled the reader to immediately and easily appreciate the location and distribution of each kind of lunar feature shown on each of the maps.

If, in writing this book, I have convinced the amateur lunar observer that Earth-based observation of the Moon is worthwhile and have helped him to derive the maximum of aesthetic pleasure from his hobby, as well as pointing the way to making contributions to knowledge, then I will be well satisfied as I will have achieved my main objective.

ACKNOWLEDGEMENTS

I wish to thank Dr Simon Mitton, Editorial Director, Science, Technical and Medical Publishing, Cambridge University Press, for his interest in my book and for expediting its production.

Many valuable suggestions were made by Mr Jeremy Smith, also of the Cambridge University Press, whom I wish to thank for his meticulous subediting of the typescript.

Dr Jean Dragesco of the Association of Lunar and Planetary Observers generously supplied several of his beautiful photographs of lunar surface features for illustrative material. The Lick Observatory authorities kindly allowed me to reproduce their photographs of the full Moon (frontispiece) and the last quarter Moon.

My thanks are due to various publishing houses for permission to redraw illustrations from their books for use as illustrative material in the text. These are:

W.W. Norton & Co. (*New Guide to the Moon*, P.A. Moore, 1976) for Figs. 1.12 and 6.14

Frederick Muller (*Our Moon*, by H.P. Wilkins, 1954) for Figs. 4.12 and 6.2

Hutchinson & Co. (*Surface of the Moon*, V. A. Firsoff, 1961) for Fig. 4.11

The Council of the British Astronomical Association allowed me to redraw illustrations from various issues of *The Moon* (BAA Lunar Section Journal) and the *Journal* of the BAA for Figs. 4.21, 4.25, 6.4, 6.10 and 6.13.

Finally, my thanks are due to my friend, Joseph Provato, for typing most of the manuscript.

F. W. Price
Buffalo, New York
USA

Introduction

Why Observe the Moon?

The best answer that anyone interested in lunar studies could give to the question posed by the title of this Introduction is: 'because I love doing it'. Here is the mark of the true amateur for whom this book is primarily written; this is someone who pursues a hobby or field of study for the sheer love of it, without any thought for material gain or notability. The amateur may, however, be sometimes incidentally rewarded by knowing that he has contributed to knowledge or advancement of his field of study as a result of his activities and researches.

Even if the work of the amateur does not contribute to knowledge or result in advancement of a field of study, this does not mean that he is wasting his time. There is much of individual educational value to be obtained from the pursuit of any branch of natural science; the mind is thereby broadened and deepened and first-hand study has an enlightening effect that is difficult to define and is definitely not attained from book study alone, important though this is. All this is a healthy antidote to the prevailing materialistic tenor of the times in which we live. Although much lip service is paid to the development and dignity of the individual, he is all too often overlooked in this technological age.

We live in times in which scientific pursuits, if they are not to be considered a waste of time, are expected to yield new knowledge or to answer hitherto unanswered questions. This is a pity because the pursuit of a scientific hobby, such as lunar observation and study for its own sake, has considerable aesthetic and other benefits to the individual as just mentioned. This is often overlooked in this present age of 'big science' with its characteristic unrelieved insistence on 'hard' data, results and publication output. The old-fashioned natural philosopher who did not care whether his work was published or not had a fund of true education, mental poise and inner contentment all too rarely found in the modern, often harassed, practitioners of 'big science'.

To come home after a tiring day at one's workplace and to view the silvery crescent of a three-day-old Moon on a cool clear spring evening with a small telescope is a most pleasant and relaxing exercise. If it lifts us out of ourselves and the workday world for only a few minutes it can only have a refreshing and exhilarating effect on our bodies and minds.

Even then, however much the amateur may enjoy the study of the Moon for its sake, there will nearly always come a time when desire to discover something new and to make a contribution to knowledge makes itself felt. At first sight there does not seem to be much hope for this today in amateur lunar observing; ever since the publication of the close-up photographs of the lunar surface obtained with the cameras aboard the lunar probes and Orbiter craft and the production of the USAF Moon maps, there has arisen not surprisingly the conviction among amateur astronomers that Earth-based lunar observation is no longer capable of yielding scientific data of permanent value. This is not the first time in the history of lunar studies that a major advance has, oddly, discouraged amateur lunar observers and made their efforts seem futile; there have been two others.

First, there was the publication in 1837 of Beer and Mädler's map of the Moon and their accompanying book. Mädler's reputation in astronomical circles was such that the map and the book were considered to be the last word on lunar matters and therefore made it unlikely that further telescopic study of the Moon would yield anything new. This feeling was reinforced by Mädler's assertion that the Moon was a dead and therefore unchanging world. This had a deadening effect on lunar studies for many years. Then, in 1866, J. Schmidt announced the 'disappearance' of the lunar crater Linné, which is discussed in detail in Chapter 6. This episode and the discovery of discrepancies between Beer and Mädler's map and charts drawn by later observers using larger telescopes made it appear that the Moon might not be a changeless world after all. The result was that telescopic observation of the Moon received a strong stimulus and lunar research was vigorously pursued in the ensuing years.

Second, there was the publication many years later of the photographs of the Moon's surface made with the 100-inch reflecting telescope at the Mount Wilson Observatory. It seemed pointless for observers using much smaller telescopes to attempt to contribute to knowledge of the Moon when such photographs became available, so that amateur lunar observers again felt discouraged. I experienced this feeling myself when I first became interested in the Moon as a schoolboy in England. The astronomy books of those days were illustrated with the Mount Wilson and other large obser-vatory photographs in the sections dealing with the Moon. I quite naturally thought that studying the Moon with much smaller instruments could never be anything more than a relaxing and educational pastime. For the next several years I almost abandoned astronomy as I was engrossed with my academic studies at university for my B.Sc. & Ph.D. degrees in biology and biochemistry. One day, I came across a copy of Wilkins and Moore's book *The Moon*, then only recently published. In it I was amazed to see illustrations so refreshingly different from the lunar photographs that were repeated in every book on astronomy I had known as a boy. Here, instead, were drawings and charts of lunar formations I had hardly heard of made with much smaller telescopes as well as drawings, not just photographs, of lunar formations made with the great 33-inch refractor of the Meudon Observatory, a telescope I never even knew existed until then. Afterwards I found other books filled with drawings of lunar surface features made by amateurs working with moderate telescopes. I quickly realised that a whole world of discovery was open to anyone with even a modest telescope who chose to study the Moon. Wilkins and Moore's book brought me back to astronomy and rekindled my boyhood passion for the Moon, which has continued to this day. To these authors I am eternally grateful.

However, as is now well known, visual observers using moderate telescopes can perceive delicate lunar surface detail in moments of excellent seeing that do not appear in photographs taken with even the largest telescopes. Further, the photographs do not show every lunar feature under all possible illumination and libration conditions so that not long afterwards, amateur lunar observation got going again. Observers used the photographs to prepare accurate outlines of lunar formations on which they would insert details seen in their visual observations that were not visible in the photographs.

Nowadays, as already mentioned, many lunar enthusiasts once more feel that it is futile to pursue cartographic observation of the Moon because of the availability of the close-up photographs of the lunar surface obtained by circumlunar orbiting space craft. Professional astronomers have now turned their attention to the Moon and have constructed accurately detailed maps of almost the entire lunar surface using the close-up photographs as a basis but, as with the Mount Wilson pictures, these modern photographs still do not show every lunar feature under every possible illumination angle and libration state and no further lunar missions are planned for the foreseeable future.

The emphasis of amateur lunar observation has now largely, but not entirely, shifted away from purely cartographic work to other areas, as described in Chapter 7. It is to be hoped that Earth-based telescopic study of the Moon will again be vigorously pursued in the coming years; there are many signs that a resurgence is already well under way.

1

Our Moon

The Moon is our Earth's nearest celestial neighbour and its only natural satellite. It is a spherical world about 2160 miles in diameter. The Earth's equatorial diameter is a little less than 8000 miles so the Moon's diameter is between one-quarter and one-third that of the Earth's (Fig 1.1). The Earth's equatorial diameter is specified here because it is somewhat greater than the diameter as measured through the poles–the polar diameter. The reason for this is that beneath the relatively thin solid crust the interior of the Earth is fluid and the Earth's axial rotation–one complete revolution taking

Fig. 1.1 Comparative sizes of the Earth and Moon

about 23 hours and 56 minutes–is fast enough to generate sufficient cen-
trifugal force to cause a slight outward bulging at the equator. The Earth
is therefore a spheroid rather than a sphere, the equatorial diameter being
7927 miles and the polar diameter 7900 miles, a difference of 27 miles.

The Moon is situated at a distance from the Earth of about a quarter of
a million miles and moves around it in an approximately circular orbit. One
complete circuit around the Earth takes somewhat less than one calendar
month.

The Earth–Moon system

The Earth is one of a system of nine worlds, or planets as they are called,
all moving in nearly circular and roughly concentric orbits around the Sun.
The Sun's family of planets is called the Solar System (Fig. 1.2). The Sun
itself is just one of the countless millions of stars that populate the Universe.
The Earth revolves around the Sun and makes one complete revolution in
its orbit in about 365¼ days (the *year* as it is called). Since the Moon revolves
around the Earth the actual path of the Moon in space is like a wavy circle
(Fig. 1.3). We have just mentioned that the Moon moves around the Earth.
This is not quite correct; the two bodies actually revolve around their common center of gravity or center of mass to be precise. Gravity is the force
that keeps planets circling around the Sun instead of wandering off into
space. It is this same force that keeps the Moon in its orbit around the Earth.

Because of the great difference in the masses of the Earth and Moon (the
ratios are Earth : Moon 81 : 1) the center of mass of the two bodies actually
lies well within the Earth and is called the *barycenter*. It is around this that
the Earth and Moon rotate. The same is true of the Earth and Sun. The
Sun is enormously more massive than the Earth and the center of mass
around which the two bodies revolve is well within the Sun.

For all ordinary purposes no great error is incurred if we consider the
Moon as revolving around the Earth or the Earth as revolving around the
Sun.

The satellites of the planets of the Solar System are all very much smaller
than the planet around which each revolves. However, as a satellite our
Moon is proportionally unusually large and massive when compared with
the Earth although in absolute terms it is average in size when compared
with some other satellites in the Solar System (Fig. 1.4). Because of this it
is probably better to consider the Earth and Moon as a double planet rather
than as a planet and satellite.

Lunar co-ordinates and quadrants

Before describing the Moon's surface features it will first be necessary to
explain the system used to indicate directions and positions on its surface.

As seen with the naked eye, the north pole of the Moon is at the top of
the disc and the south pole at the bottom. When viewed with an astronomical telescope, the Moon's image appears inverted, as does that of any object
as explained in Chapter 3. Hence, all lunar maps, charts and drawings

made before the 'Space Age' show the Moon inverted with south at the top and north at the bottom, for ease of comparison with the telescopic image. Again, as seen with the naked eye, the right hand limb of the Moon is closer to the observer's western horizon and the left hand limb is closer to the eastern horizon. Hence, ever since telescopic observation of the Moon began, the Moon being considered a purely celestial object, the 'right'

Fig. 1.2 The Solar System

Orbits of the Outer Planets

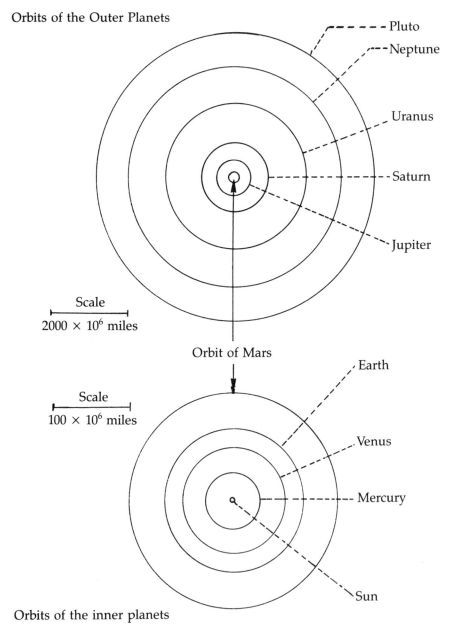

Orbits of the inner planets

Fig. 1.3 Path of the Moon in space (not to scale)

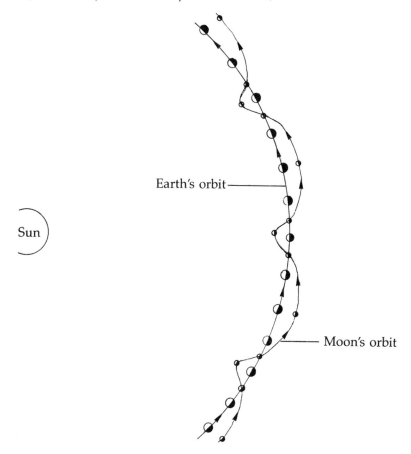

Earth's orbit————

Sun

Moon's orbit

Fig. 1.4 Comparative sizes of our Moon, satellites of Jupiter (above) and satellites of Saturn (below).

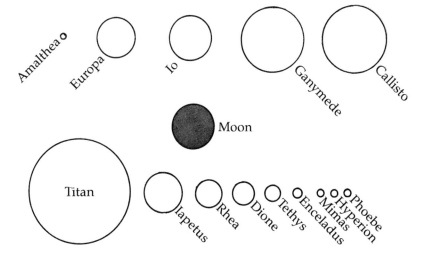

Amalthea

Europa

Io

Ganymede

Callisto

Moon

Titan

Iapetus

Rhea

Dione

Tethys

Enceladus

Mimas

Hyperion

Phoebe

and 'left' limbs of the Moon have customarily been designated the west and east limbs, respectively. In the inverted telescopic image of the Moon the four cardinal points are therefore arranged as in Fig. 1.5. This illustration also shows the traditional division of the Moon's disc into four quadrants; beginning with the north-west quadrant they are numbered 1–4 in a counterclockwise direction.

The time-honored convention of designating the naked eye 'right' and 'left' limbs of the Moon as west and east is the reverse of the west and east directions used for the Earth so that to an observer on the Moon the Sun would rise in the west and set in the east.

The first lunar astronauts decided that they couldn't get used to all this so the matter was debated by the International Astronomical Union (IAU). It was decided overwhelmingly to reverse the centuries-old convention of designating east and west on the Moon and this was made official in 1961. Soon afterwards, lunar maps and charts began to appear with the Moon 'right way up' with north at the top and with the east and west limbs in the same positions as they are in terrestrial maps, i.e, the reverse of what

Fig. 1.5 Lunar cardinal points (classical, with inverted telescopic image)

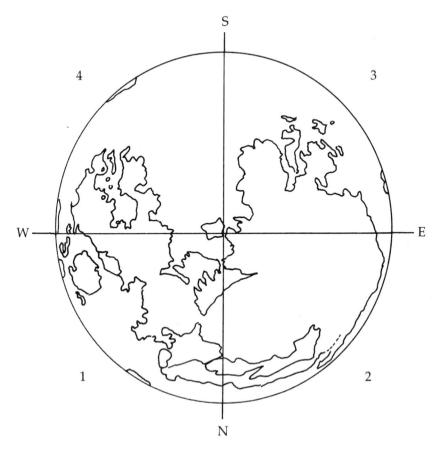

they were in all previous lunar literature, charts and maps (Fig. 1.6). This creates a problem for the practical lunar observer when consulting lunar literature and charts because in the vast bulk of it which is most likely to interest the observer the older (classical) cardinal points and east–west directions are used. To avoid confusion, especially that caused by reversal of the east and west directions in the IAU convention, I have decided to use the classical lunar coordinates throughout this book so that descriptions of the positions of lunar surface features, especially where the terms 'east' and 'west' are used, will be consistent with the similar usage in the literature most likely to be of interest to the practical observer. The lunar student will therefore be relieved of the necessity of constant mental switching of east and west when studying this literature as would be necessary if I had decided to use the IAU convention. Apart from anything else, the IAU system of coordinates gives rise to anomalies in nomenclature; for example, the so-called Mare Orientale (Eastern Sea) is now in the western hemisphere of the Moon!

Fig. 1.6 Lunar cardinal points (IAU convention)

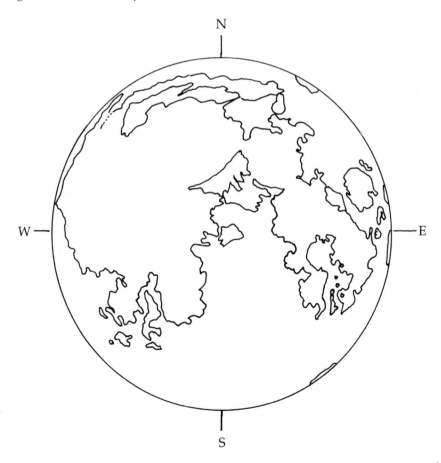

The Moon's surface features

Many amateur astronomers are understandably fascinated by the Universe as a whole and delight in observing distant galaxies, nebulae, star clusters and other remote stellar objects. They either make for themselves or purchase the largest practicable telescopes and seek the darkest skies away from the light pollution of cities. There is a friendly rivalry among them as to who can 'pick up' this or that very faint or elusive hazy spot (galaxy or nebula, usually) in the sky. One of these 'deep sky' enthusiasts once facetiously remarked to me that 'the Moon is so boring, it is only a great ball of rock'. One is entitled to one's views, of course, but that mere 'ball of rock' is our nearest neighbor in space and has a special fascination of its own to many astronomers which is different from that of 'deep sky' objects. It is to these folk that this book is dedicated. In the words of Rev. T.W. Webb, 'Many a pleasant hour awaits the student in these wonderful regions'.

Until the advent of the rocket-propelled American and Russian lunar orbiting and landing vehicles and the 'Space Age', all our knowledge of the Moon's surface features was obtained entirely from Earth-based telescopic study. On July 31, 1964 the American Ranger 7 space vehicle hit the Moon's surface and destroyed itself, but during the preceding few minutes it transmitted to Earth over four thousand excellent photographs and the Moon's surface could be studied at really close range for the first time. The greatest contribution of the space program to lunar specialists interested in regional mapping was the series of five Lunar Orbiter photographic missions. These were completed in a year between August 1966 and August 1967. Orbiters I–III circled the Moon in nearly equatorial orbits so as to photograph possible landing sites for the Apollo space craft. Orbiter IV in a near-polar orbit secured pictures of the entire Earthward hemisphere and Orbiter V, also in near-polar orbit, gathered pictures of certain selected areas on the Earthward side and completed the coverage of the averted side. The total number of pictures obtained was 1950. These pictures reveal fine detail never before seen with Earth-based telescopes. As a result, it seemed that further telescopic study of the Moon from the Earth would be pointless, but there is no need for the amateur lunar observer to be discouraged, for the reasons mentioned in the Introduction.

In the following description of the Moon's surface features, frequent references will be made to the results of close-up lunar photography although our main concern will be with telescopic work in the rest of this book.

Our satellite has virtually no atmosphere and no liquid water has ever been detected on its surface so that the Moon is uninhabited and life as we know it could not exist there. Apart from the absence of oxygen implied by the absence of air the lack of an atmospheric mantle and clouds exposes the Moon's surface to extremes of temperature unknown on the Earth that virtually no known living thing could survive. The absence of atmosphere and clouds is, however, an enormous advantage for the observer of the Moon because the Moon's surface features will therefore not be obscured or rendered indistinct. In fact, the surface is always seen with startling sharpness and clarity.

A casual naked eye glance at the full Moon will show that the whitish disc is mottled with dusky greyish spots and patches. To people with a lively imagination the total configuration of these markings bears a resemblance to a human face, the well-known 'Man in the Moon'. Other smaller groups have been likened to a man carrying a bundle of sticks, a woman reading a book and the profile of a woman kissing someone! Other imaginary figures are the crab and the hare.

A small telescope or even a pair of binoculars shows that these greyish areas are sprinkled with bright spots, patches and streaks as are also the light-colored areas. A prominent feature of the full Moon is the system of bright white streaks radiating from a point near the south limb. A less spectacular radial streak system is situated on one of the large dark areas near the east limb.

The general telescopic appearance of the full Moon is rather flat like a map without any surface relief and everything is dazzlingly brilliant. This is because the Sun is shining straight down on to the Moon as seen from the Earth and its light is reflected directly into our eyes. At other phases the solar illumination is oblique and shows the Moon's surface to be very rugged. Strong shadows are cast and this is especially noticeable near the *terminator*, where the Sun is rising and the illumination angle is extremely oblique. The lighter areas are the most rugged but the dusky areas are relatively smooth. In a small telescope the most eye-catching surface features are the innumerable ring-like formations usually called *craters*, peppered over the entire lunar surface. They are most numerous in the light-colored regions where in some places they literally jostle each other but they are much less numerous in the greyish areas. When close to the terminator, the prominent shadows within the craters and those cast by their raised walls give the impression that they are deep cup-like depressions scarring the Moon's surface. This is an illusion caused by the extremely oblique illumination. The profiles of the craters are more like very shallow saucers than bowls or cups (Fig. 1.7).

Craters are found in all sizes ranging from huge enclosures well over 100 miles across to pits well below the resolution of the largest Earth-bound telescopes, as revealed by lunar close-up photographs obtained with orbiting space vehicles.

We will now survey the different types of lunar surface features and their distribution.

Maria ('seas'). The dusky grey patches that mottle the bright disc of the full Moon were thought by the ancients to be seas and have borne that designation ever since although it has been known for a long time that they are not bodies of water; in fact, there is no appreciable accumulation of liquid water anywhere on the Moon's surface. The ancients didn't seem to realise that if the maria were really seas they would have seen the Sun brilliantly reflected in them. The 'seas' are usually called by their Latin names of *Maria* (singular *Mare*) and have been given fanciful names such as the Mare Crisium (Sea of Crises), Mare Imbrium (Sea of Showers) and Oceanus Procellarum (Ocean of Storms) (Fig. 1.8). Smaller areas of mare-like

11

Fig. 1.7 Sectional profiles of some large lunar ring structures (redrawn after W. Goodacre)

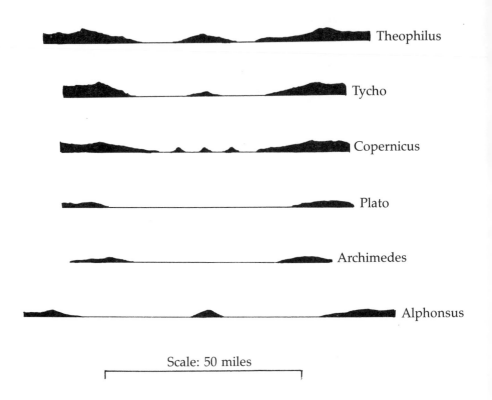

Scale: 50 miles

material are called lakes such as the Lacus Somniorum (Lake of Dreams) or marshes such as the Palus Somnii (Marsh of Sleep). Small detached patches of dark marial material are found singly or in groups scattered in various parts of the surface, often on the floors of some of the larger craters.

Many of the maria approximate closely to a circular shape such as the Mare Imbrium in the second quadrant and the Mare Serenitatis in the first quadrant. Others are somewhat rectangular such as the Mare Nectaris, Mare Foecunditatis and Mare Nubium. Yet others are quite irregular like the vast Oceanus Procellarum or are elongated like the Mare Frigoris.

The boundaries of the maria are indented in many places by roughly semicircular 'bays', the largest and best known of these being the Sinus Iridum (Bay of Rainbows) on the north-east 'shore' of the Mare Imbrium.

In addition to the larger maria there are several minor maria most of which are close to the limb regions. On the west limb, proceeding from north to south, are the Mare Humboldtianum at about latitude 60° north, Mare Novum, Mare Marginis (opposite Mare Crisium), Mare Smythii on the west point of the west limb and Mare Australe at about latitude 50° south. On the east limb are the two closely apposed Maria Veris and Autumnii between about latitudes 10° and 15° south. These are actually patches of mare material between the mountain rings of the Mare Orientale, shown

Fig. 1.8 Marial bodies

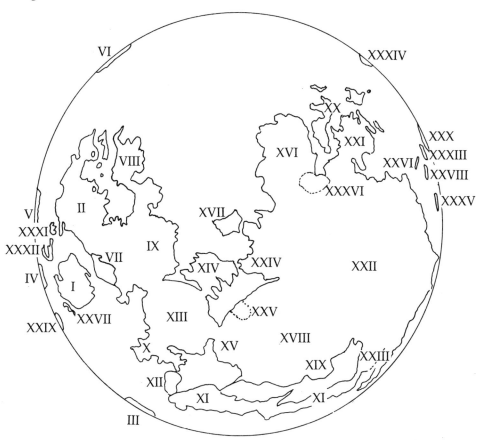

Key

Maria ('Seas')

I. Mare Crisium
II. Mare Foecunditatis
III. Mare Humboldtianum
IV. Mare Marginis
V. Mare Smythii
VI. Mare Australe
VIII. Mare Nectaris
IX. Mare Tranquillitatis
XI. Mare Frigoris
XIII. Mare Serenitatis
XIV. Mare Vaporum
XVI. Mare Nubium
XVIII. Mare Imbrium
XXI. Mare Humorum
XXII. Oceanus Procellarum
XXVI. Mare Aestatis
XXVII. Mare Anguis
XXVIII. Mare Autumnis
XXIX. Mare Novum
XXX. Mare Orientale
XXXI. Mare Spumans

XXXII. Mare Undarum
XXXIII. Mare Veris
XXXIV. Mare Parvum
XXXV. Mare Hiemis
XXXVI. Mare Cognitum

Lacus ('Lakes') *and Paludes* ('Marshes')

VII. Palus Somnii
X. Lacus Somniorum
XII. Lacus Mortis
XV. Palus Nebularum
XX. Palus Epidemiarum
XXV. Palus Putredinus

Sinus ('Bays')

XXVII. Sinus Medii
XIX. Sinus Iridum
XXIII. Sinus Roris
XXIV. Sinus Aestuum

in 'full face' view in the Lunar Orbiter IV frame 187M. Close to the west limb and to the west of Mare Foecunditatis are the two small irregular Maria Undarum and Spumans and on the north-west 'shore' of Mare Crisium abuts the Mare Anguis. These three appear to be groups of coalesced ring structures that have become flooded with mare material.

Thirty-two of these marial bodies have been given separate names. They occupy a large fraction of the Moon's Earthward face and are concentrated in the first three quadrants. They predominate in the northern hemisphere. The only area that is almost devoid of them is the sector bounded by the center of the Moon's disc and the south and south-west points of the limb.

The dark material of the marial surfaces is often called *lunabase* and the brighter material of the upland areas is likewise called *lunarite*. The major maria form an interconnecting system, the only prominent isolated example being the Mare Crisium. Other smaller isolated maria 'hug' the limb such as the Maria Humboldtianum, Australe and Marginis.

In small telescopes the maria appear to be quite smooth plains. Larger telescopes show them to be not perfectly smooth and, under low solar illumination angles, gentle undulations, low blister-like swellings and other surface irregularities are seen. The Soviet Luna IX space vehicle was launched on January 13, 1966 and was the first to make a soft landing on the Moon's surface. It came down in the Oceanus Procellarum about 44 miles north-east of the ring structure Cavalerius. Photographs taken from it show the marial surface to be rough and coarsely porous with cavities about 1 cm in size. Marial surface material appears to have the mechanical strength of wet sand as was indicated by the Surveyor III space craft digging activities and the impact of the landing pads of Surveyor I.

Some of the maria are deeper than others and so their surfaces are not part of the same sphere. A few such as the Maria Crisium, Nectaris and Imbrium are partly or completely surrounded by huge mountain walls.

The maria are not of a uniform grey color, some being darker than others. Considerable regional variations of intensity are noted in individual maria, which may vary from a light yellowish grey to deep grey. To a certain extent these differences depend on the solar illumination angle. Indistinct colors have been seen in some, such as the yellowish−green tint in Mare Frigoris and Mare Imbrium. The Mare Serenitatis and Mare Crisium frequently exhibit a greenish hue under a high Sun. Elsewhere, decided sepia colors have been detected under a low Sun. Generally speaking, the colors are difficult to see and require excellent seeing conditions for successful detection. The subject of lunar colors is dealt with in much more detail in Chapter 7.

Ring structures (Fig. 1.9). The ring structures, or *craters* as they are collectively named, are probably the most interesting of lunar surface features. Their bewildering number and variety of form have fascinated lunar observers ever since telescopic observation of the Moon began. It is to them more than anything else that the Moon's characteristic pock-marked surface appearance in the telescope is due and which never fails to enthrall the observer, no matter how familiar the Moon's surface may become.

Individual ring structures are named after famous astronomers, scientists,

Fig. 1.9 Major ring structures (55 miles and over)

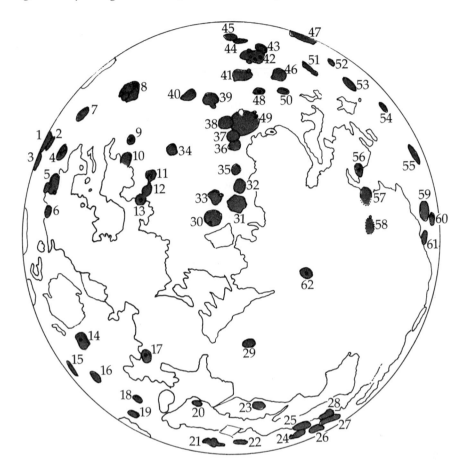

Key

1.	Wilhelm Humboldt	22.	Goldschmidt
2.	Phillips	23.	Plato
3.	Hecataeus	24.	Anaximander
4.	Petavius	25.	John Herschel
5.	Vendelinus	26.	Pythagoras
6.	Langrenus	27.	Babbage
7.	Furnerius	28.	South
8.	Janssen	29.	Archimedes
9.	Piccolomini	30.	Hipparchus
10.	Fracastorius	31.	Albategnius
11.	Catharina	32.	Alphonsus
12.	Cyrillus	33.	Ptolemaus
13.	Theophilus	34.	Sacrobosco
14.	Cleomedes	35.	Arzachel
15.	Gauss	36.	Purbach
16.	Messala	37.	Regiomontanus
17.	Posidonius	38.	Walter
18.	Atlas	39.	Stöfler
19.	Endymion	40.	Maurolycus
20.	Aristoteles	41.	Maginus
21.	Meton	42.	Clavius

43.	Blancanus
44.	Gruemberger
45.	Moretus
46.	Longomontanus
47.	Bailly
48.	Tycho
49.	Deslandres
50.	Wilhelm I
51.	Schiller
52.	Phocylides
53.	Schickard
54.	Lagrange
35.	Darwin
56.	Gassendi
57.	Letronne
58.	Ruined ring north of Flamsteed
59.	Grimaldi
60.	Hevelius
61.	Copernicus
62.	

mathematicians, philosophers and other notable personalities of the past, such as Eratosthenes, Archimedes, Plato, Julius Caesar, Mädler and Schröter. Small rings close to the larger ring structures are often named after the larger ring followed by a capital letter, e.g., Posidonius J and Archimedes A.

Attempts have been made to classify the ring structures according to their morphology, size and other characteristics. Although not entirely successful their grouping into different classes with distinctive names has been helpful to the practical observer. In addition, these groupings probably correspond to actual differences in origin. Regarding the origin of lunar ring structures, there are two main opposing hypotheses. One attributes their formation to random impact of meteoroidal bodies on the Moon's surface in the remote past. This is the so-called impact, or meteoritic or exogenic hypothesis. Some authorities believe that even the maria were formed by impact of large bodies of asteroidal dimensions. Superficially, the random impact hypothesis has many attractive features but there are serious objections to it on both theoretical and observational grounds. The other postulated mechanism supposes that lunar ring structures were formed as the result of forces from within the Moon itself, the so-called igneous, or volcanic or endogenic hypothesis. The term volcanic is misleading since lunar ring structures are nothing like terrestrial volcanoes in appearance.

Discussion of the origin and nature of the ring structures and of the meteoritic versus volcanic hypothesis is beyond the scope and purpose of this book but readers who wish to pursue this aspect of lunar study further will find a list of books dealing with the origin of lunar ring structures at the end of this chapter.

Lunar ring structures are classified as follows.

i *Walled plains.* These objects are approximately circular in outline but some of the largest are roughly rectangular or polygonal. They are from about 180 to 60 miles or less in diameter and are bordered by massive, broad and complex ramparts. The rampart is usually cut through by numerous gaps and valleys or is encroached upon by smaller ring structures, presumably of more recent date. The enclosed plain or floor is usually relatively smooth and usually only slightly depressed below the level of the surrounding surface or less often is considerably depressed. In the case of the formations Mersenius and Petavius, the floor is convex. The floor is often of a dusky tint and generally has hill-like elevations, ridges and craters upon it. Linear groups of numerous small craters which are sometimes prolonged by clefts or cracks in the surface are often found. Whitish streaks may traverse the floor as in the case of Archimedes and Plato. Occasionally, it may be cut through by deep furrows known as rills or clefts.

The walled plains are mostly found in the south-east and south-west quadrants. A good example is Ptolemaus, which is 115 miles in diameter with an almost perfectly hexagonal outline. Its rampart is interrupted by valleys and passes in many places and one peak reaches a height of 9000 feet above the level interior plain, which is at about the same general level of the surrounding terrain. Under very low solar illumination the floor of

Ptolemaus is seen to be sprinkled with many shallow saucer-like depressions. The Apollo 16–0579 photograph of Ptolemaus shows the floor to be thickly peppered with myriads of tiny pits well beyond the resolution of Earth-based telescopes.

Near to Ptolemaus is the walled plain Alphonsus, which is somewhat smaller. Its floor is variegated with dark spots of marial material, three of which are large and prominent. There are also some clefts. The Ranger 9 space craft landed inside Alphonsus and the close-up views of these clefts show that they are actually chains of craterlets that have run together. A ridge runs across the floor in a roughly north to south direction and upon it stands a central mountain. Other examples of walled plains in this region are Arzachel (diameter 60 miles), to the south of Alphonsus, and Albategnius (diameter 80 miles), which lies to the west of Alphonsus and Ptolemaus.

One of the best-known walled plains is the splendid formation Clavius, 145 miles in diameter (Fig. 1.10). It has a floor depressed well below the level of the surrounding surface and upon it is a fine curved group of four smaller rings that progressively decrease in size from one end to the other. Like all other similar formations Clavius is a magnificent spectacle at lunar sunrise.

Fig. 1.10 Clavius (from a photograph) F.W. Price

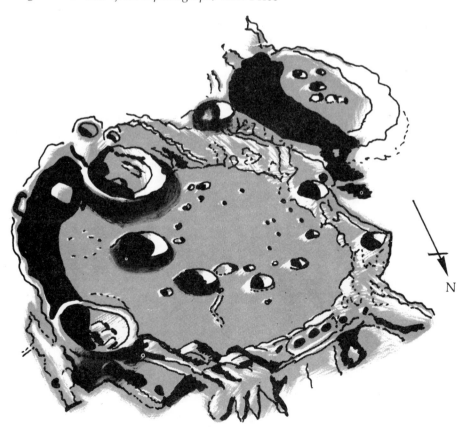

N

The largest of the walled plains is Bailly, which is about 180 miles in diameter and situated on the limb exactly between the south and south-east points. This formation is so large that had it been nearer the center of the disc it would no doubt have been classified as a mare. The same may be said of the great dark-floored walled plain Grimaldi (diameter 120 miles) close to the Moon's east limb, and Schickard (diameter 134 miles) near the south-east limb (Fig. 1.11). In fact it is difficult to know where to draw the line between a small mare and a large walled plain. The walled plains have a pronounced tendency to be aligned in meridional chains (fig. 1.12). Running down the center of the moon's disc in the southern hemisphere is the great longitudinal chain of walled plains consisting of Walter, Regiomontanus, Purbach, Arzachel, Alphonsus and Ptolemaus. Close to the east limb is the great chain of formations Grimaldi, Hevelius and Cavalerius and running close to the west limb is the Furnerius, Petavius, Vendelinus and Langrenus chain of walled plains. The Mare Crisium probably belongs in this chain although it is much larger than the other members. Non-random alignment of walled plains has implications regarding their origins and points to an endogenous rather than a random impact mechanism.

Among the walled plains the 55-mile Wargentin, which lies just to the south-east of Schickard, is almost unique in that its floor is raised well above the surrounding surface so that under the right lighting conditions it looks rather like an oval slab of stone on the Moon's surface. It seems that at some remote epoch, the interior became filled with molten lava to the brim and later solidified. Related to the walled plains are some fairly prominent roughly circular areas with rather low walls that are often only slightly smaller than the largest walled plains. An example is Deslandres, about 100 miles across and more rectangular than circular in outline, adjacent to the east wall of Walter. Its floor is slightly lower than the neighbouring Mare Nubium. The entire floor of Deslandres is covered with numberless craters and crater pits and its south wall is encroached upon by the ring structure Lexell. On the east side of the floor is the ring structure Hell. Deslandres is sometimes called Hörbiger or Hell Plain. A less well marked

Fig. 1.11 Schickard. May 27, 1961. 20.55–23.15 BST. F.W. Price. Three-inch refractor

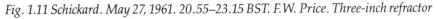

N

example lies to the north of Deslandres and is noted for the great fault known as the 'Straight Wall' or the 'Railway' that traverses its floor in a roughly north to south direction.

ii *Mountain rings.* These vary in size from about 70 miles in diameter to 15 miles or smaller. Structurally, they appear to be related to the walled plains but present a ruinous appearance probably due to their having been partially inundated with marial material at some remote time. A typical mountain ring consists of a roughly circular discontinuous rampart made up of isolated blocks, rarely more than a few hundred feet in height. Others are formed from discontinuous curved wall sections joining up to form a roughly circular formation.

The mountain rings are mostly found on the maria; the Oceanus Procel-larum and Mare Imbrium contain numerous specimens. Examples are the

Fig. 1.12 Meridional alignment of large walled plains (Redrawn from P.A. Moore, *New Guide to the Moon*, Norton, 1976, by permission. Copyright © by P.A. Moore).

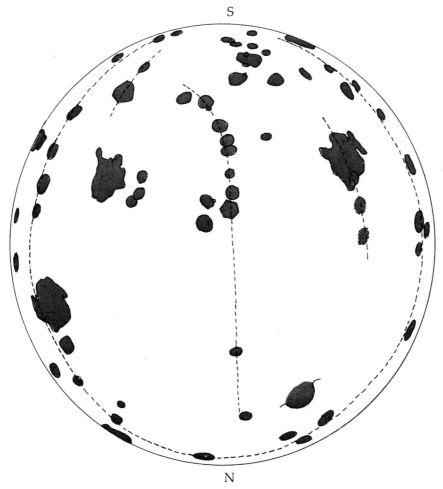

S

N

incomplete ring of Letronne, about 60 miles across, in the south Oceanus Procellarum, the ring south-east of Encke and a third lying to the west of Letronne north of Wichmann. The formation in the Mare Imbrium immediately south of Plato is another example and was named Newton by Schröter. Sometimes mountain rings are found on the floors of walled plains such as the ring on the north part of the floor of Catharina.

iii *Ring plains.* These are smaller and far more numerous than the walled plains or mountain rings. They are from 60 miles to 10 miles in diameter. The ramparts are more continuous than those of walled plains and are circular and internally slope down quite steeply to the floor, which is well depressed below the level of the outer surface. The ramparts are broken in places by gaps, craters and other depressions. An interesting and suggestive structural feature of some ring plains is that gaps and discontinuities are found on the north or south points of the rampart or on both, but nowhere else.

The interior walls of ring plains often exhibit prominent, more or less concentric terracing which makes them especially fascinating objects under appropriate solar illumination and with high telescopic powers. On the floors of the ring plains are often found a central mountain mass which may be a single peak, or more often there are multiple peaks or blocks as in Bullialdus and Aristillus. Some central mountains are quite small in comparison to the size of the ring plain whereas others may be very large and occupy most of the floor, such as the enormous central mountain in Alpetragius. Occasionally, a crater is found instead of a central mountain, as in Posidonius and Hesiodus.

Not infrequently, the ring plains are found in north–south aligned pairs such as Aristillus and Autolycus in the Mare Imbrium, Eudoxus and Aristoteles to the south of the west extremity of Mare Frigoris and Godin and Agrippa near the center of the Moon's disc. In each case, the northern member of each pair is larger than the southern member.

The most splendid example of a lunar ring plain is undoubtedly Copernicus. It stands on an elevated area of the lunar surface between the Mare Imbrium and Mare Nubium. The massive walls are 56 miles across and are surmounted by 50 or more peaks. The floor is about 40 miles across and on it stands a multiple mountain mass made up of seven separate blocks. The outer walls are very complex and marked by both radial and concentric ridges. Vast numbers of hillocks and tiny craterlets appear where the radial ridges taper to fine points. As the Sun rises, a system of wispy bright rays develops radiating from Copernicus across the surrounding surface.

iv *Craters.* Under this heading are an enormous number of circular or near-circular formations that vary in diameter from 30 miles to less than five miles. Their narrow walls rise more steeply from the outside than those of the ring plains and internally descend to relatively deep hollows. Many have small central peaks that are sometimes difficult to see except in large telescopes. Craters often have a 'fresh' appearance as though they were relatively young formations and often are quite bright. Some stand on bright areas. They are found in all situations—on the maria, in the uplands and on the walls or floors of walled plains. They frequently intrude upon and overlap one another, the smaller ones almost invariably overlapping the larger. An interesting example is the crater Thebit on the west 'shore' of the Mare Nubium, which has a smaller crater intruding on its north–east wall, which in its turn has a yet smaller crater overlapping its north–east wall. This invariable overlapping of large formations by smaller also applies to ring structures other than craters.

Craters often form non-randomly arranged groups, as do many of the walled plains. They may form repetitive linear patterns where the individual craters have similar morphology (consanguineous progressions) as in the Tacitus–Aliacensis group, decremental arcuate progressions where the individual craters consecutively decrease in size as on the floor of Clavius and near the craters Halley and Condamine, or rhomboidal and triangular patterns as are found on the Mare Imbrium.

v *Craterlets*. These are simply very small craters and are found in vast numbers. They have no visible central peaks. Many are found on the floors of large walled formations such as Archimedes, Plato and Goldschmidt. They often occur in long chains. The smaller craterlets are good tests of telescopic quality.

vi *Crater cones*. These resemble terrestrial volcanoes. They are conical peaks with cup-shaped or conical depressions on the peak. In most cases the central depressions are detectable only in giant telescopes. They are abundant near the formation Stadius and two notable specimens stand near to Bailly. Under high light they appear as white spots.

vii *Crater pits*. Different from all the above are crater pits, small quite shallow depressions without raised rims. Often they are irregular in shape. There are enormous numbers of them and usually they are visible only under very oblique solar illumination. Most require considerable telescopic power to detect them. The outer walls of some of the larger ring plains are peppered with them, a good example being the outer wall of the ring C.F.O. Smith (formerly Vendelinus C). Crater pits are especially abundant in the Moon's fourth (south–west) quadrant.

Shallow pits and depressions only a few inches in diameter, well below the resolution of the largest Earthbased telescopes, were photographically recorded by the American Ranger 7 space vehicle less than a second before it crash-landed in the Mare Nubium near the formation Guericke.

viii *Obscure or elementary rings*. These objects are mainly found on the surface of the maria. There is no central depression and the walls are quite low and simple in structure. They have often been interpreted as being the eroded remains of once-prominent structures but G. Fielder, who called them 'elementary rings', considers that at least some of them are incomplete or partly formed rings. The force that initiated them presumably died down before they were fully formed.

21

ix *Ghost rings.* These are bright ring-shaped features that have no raised walls and are visible only by contrast with the darker surrounding country. They are found on the maria and may be craters that became submerged beneath molten marial material in the remote past. They are most prominent under highlight conditions.

Mountains and mountain ranges. There are several fine mountain ranges on the Moon, most of them named after terrestrial ranges (Fig. 1.13). Lunar mountain ranges bear only a superficial resemblance to terrestrial ranges. Whereas our Earthly Alps and Apennines were shaped by crustal folding and later modified by water and ice erosion, the lunar mountain ranges consist of tilted crustal blocks that have been shattered by tectonic movements.

The south, west and north boundaries of the Mare Imbrium are defined by several mountain ranges so that this mare has very much the appearance of a giant walled plain. Of these mountain ranges the most splendid are the Appenines; indeed they are the finest range on the Moon's visible face. They extend in a continuous gentle curve for a distance of 400 miles from just north of Eratosthenes at their southern extremity to a point some 60 miles south–west of the ring plain Autolycus. Their general trending direction is south–east to north–west. The Apennines present a steep scarp face to the Mare Imbrium and a very broad gentle slope away from it in the direction of Mare Vaporum. There are about 3000 separate peaks in the Apennines. Among the most noteworthy of these is Mount Hadley, near the northern end of the range, which rises to a height of 15 000 feet. The highest peak is Mount Huygens at latitude 20° north, which towers to over 18 000 feet above the plain. Another peak is Mount Bradley (14 000 feet), roughly half way between Mounts Hadley and Huygens. Near the southern end of the range is Mount Wolff, rising to 12 000 feet above the plain.

The broad gentle slope of the Apennines facing away from Mare Imbrium is 160 miles wide from east to west in places. It is scored all over by intricate winding valleys that generally trend towards the south–west. In places there are mountain rings and bright craters.

The Caucasus Mountains which form part of the west wall of Mare Imbrium are a somewhat triangular range, the sharpest point of the triangle pointing southwards across the 'channel' connecting Mare Imbrium and Mare Serenitatis, to the north end of the Apennine range. To the east of the crater Calippus, at the north end of the Caucasus, is a peak towering to 19 000 feet.

Between the Caucasus Mountains and the walled plain Plato are the lunar Alps, which form a north–west wall to the Mare Imbrium. This range contains several hundred peaks; of these, Mont Blanc rises to 12 000 feet. West of Plato is another peak rising to almost the same height as Mont Blanc. Other peaks from 5000 to 8000 feet are frequent. Perhaps the best known feature of the Alps is the great Alpine Valley that cuts straight through the range in a south–east to north–west direction west from Plato. The Lunar Orbiter V frame 102M shows a sinuous rill running the length

Fig. 1.13 Mountain ranges and isolated mountains

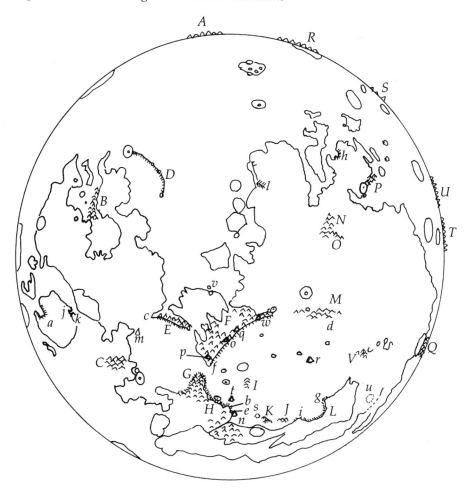

Key

Mountain ranges

H. Alps
D. Altai scarp
F. Apennines
M. Carpathian Mts
G. Caucasus Mts
U. Cordillera Mts
T. D'Alembert Mts
R. Doerfel Mts
E. Haemus Mts
V. Harbinger Mts
Q. Hercynian Mts
L. Jura Mts
A. Leibnitz Mts
P. Percy Mts
B. Pyrenee Mts
N. Riphaen Mts

S. Rook Mts
I. Spitzbergen Mts
J. Straight Range
C. Taurus Mts
K. Teneriffe Mts
O. Ural Mts

Capes and promontories

a. Prom. Agarum
b. Prom. Agassiz
c. Prom. Acherusia
d. Cape Banat
e. Prom. Deville
f. Cape Fresnel
g. Prom. Heraclides
h. Cape Kelvin
i. Prom. Laplace

j. Prom. Lavinium
k. Prom. Olivium
l. Prom. Taenarium

Peaks and isolated mountains

m. Mt Argaeus
n. Mt Blanc
o. Mt Bradley
p. Mt Hadley
q. Mt Huygens
r. Mt La Hire
s. Mt Pico
t. Mt Piton
u. Mt Rümker
v. Mt Schneckenberg
w. Mt Wolff

of the Alpine Valley down the centre of the floor, quite beyond the reach of ordinary Earth-based telescopes. Continuing eastwards from Plato we come to the highlands bordering Sinus Iridum, the previously mentioned 'bay' on the north–east 'shore' of Mare Imbrium. These highlands are named the Jura Mountains and are among the loftiest and most complex on the moon's surface. Like the Apennines the Jura show a steep scarp towards the mare and a broad gentle slope away from it. Unlike the Apennines the Jura Mountains have not been shattered and form a compact scarp. Of the many peaks in these mountains, the highest near the ring plain Sharp attains a height of 15 000 feet.

Travelling due south across the vast expanse of the Mare Imbrium to its opposite side brings us to the Carpathian Mountains forming the southern part of the mountainous border of the Imbrian Plain. They extend in a westerly direction for a distance of 180 miles from an area some distance north of Copernicus to the crater Tobias Mayer. They have a 'broken up' appearance and provide a vivid example of the shattered crustal block structure of lunar mountains. They are a fine telescopic spectacle at lunar sunrise. The highest peak, north–west of Tobias Mayer, rises to 7000 feet.

The Taurus Mountains are an irregular region of block mountains rather than a well-defined range, lying to the west of Mare Serenitatis and extending towards the craters Berzelius and Geminus in a south–westerly direction. They form a border to the west side of the Lacus Somniorum. Near to Berzelius are peaks rising to at least 10 000 feet.

On the opposite side of the Mare Serenitatis are the Haemus Mountains, which form a border to part of the south–east 'shore' of the mare. At their north–east extremity they coalesce with the Apennines to the north–west of the Mare Vaporum. On the Moon's north–east limb to the east of the walled formation Otto Struve are the Hercynian Mountains. Their nearness to the limb makes them difficult to study. There are abundant lofty peaks.

The Harbinger Mountains are situated to the north–east of the crater Aristarchus. Some of the peaks here are as high as 7000 feet.

Some smaller ranges are the Teneriffe range on the Mare Imbrium to the south-east of Plato and due east of these the Straight Range, which is remarkably linear, hence its name. It is oriented in an east–west direction. The Spitzbergen Mountains, also in the Mare Imbrium, are situated north of the walled plain Archimedes.

As well as mountain ranges there are several isolated peaks on various parts of the Moon's surface. Of these, the best known are Pico, a splendid isolated mountain in Mare Imbrium south of Plato, rising about 8000 feet above the plain; Piton, 7000 feet high, an isolated mountain on the Mare Imbrium east of Cassini, and LaHire, again in the Mare Imbrium, a large bright isolated mountain north–east from Lambert and rising to nearly 5000 feet.

There are not as many or as prominent mountain ranges in the Moon's southern hemisphere apart from those on the limb. The Altai Mountains or scarp form a splendid 275-mile curved range concentric with the margin of Mare Nectaris extending in a north–east direction from the east wall of Piccolomini to a point west of Fermat; here they take a more northerly direction and end between Catharina and Tacitus. The highest peak towards the middle of the range attains a height of 13 000 feet.

The Pyrenee Mountains are a meridionally aligned range on the west of Mare Foecunditatis and are about 190 miles long. They are sliced through in several places by valleys. One of the peaks to the east of Guttenberg has a height of nearly 12 000 feet.

In the Mare Imbrium are the Riphaen Mountains to the south of Lansberg. These are a scattered group the highest of which attains a height of only 3000 feet. They are remarkably bright, are aligned in a meridional direction and are about 100 miles long. They consist mostly of long branching arms extending to the north and south from an area west of Euclides. The short Ural range is immediately to the south of the Riphaens. Both the Riphaens and the Urals have the appearance of a group of ruined and partly submerged ring structures. The Percy Mountains are bright highlands reaching easterly from the ring plain Gassendi towards Mersenius and thus form a north–east border to the Mare Humorum. The most prominent feature is the long mountain arm that projects from the east wall of Gassendi for a distance of 50 miles in a south–east direction. There are several mountain ranges on the lunar limb. Close to the south pole are the Leibnitz Mountains extending along the limb between south latitude 80° on the west to a similar latitude on the east thus covering an arc of more than 30°. There are enormous peaks, some as high as 26 000 feet or more, in the Leibnitz range and plateaus which seen in profile on the limb make a striking telescopic spectacle. Some parts of the range appear to be profile views of large ring structures. We see the mountains on the limb regions almost as we would see them on Earth, from a near-horizontal viewpoint instead of the 'overhead' view we have of the Apennines, for example. The best views of the Leibnitz range are obtained at a favourable libration (see Chapter 2) when the Moon is about 3–4 days old. Often, the cusp of the crescent Moon appears to be prolonged by the illuminated mountains and isolated peaks shine like stars in the blackness beyond the cusp. From about latitude 80° to 57° on the east limb are the Doerfel Mountains with peaks exceeding 20 000 feet in height. According to Schröter, there are three peaks higher than 26 000 feet.

Between south latitudes 35° and 18° on the east limb are the Rook Mountains. Some of the peaks rise to about 20 000 feet. If Schröter is correct, some peaks are as high as 25 000 feet.

The Cordillera Mountains are found between south latitudes 20° and 10°, on the east limb to the east of Crüger. There are peaks over 18 000 feet high. One of the highest is situated south–east of the ring formation Rocca The Cordillera Mountains and Rook Mountains are actually the outer and inner rings respectively of the multi-ring structure known as the Mare Orientale. This is situated right on the east limb and is only ever seen in extreme foreshortening from the Earth. The Lunar Orbiter IV frame 187M is a magnificent 'full face' picture of this extraordinary formation.

The D'Alembert Mountains are situated to the east of Rocca and Grimaldi and some of the peaks are nearly 20 000 feet high. East of Riccioli are three peaks and a curious table mountain somewhat more northerly.

Wrinkle ridges (Fig. 1.14). These are peculiar to the maria and look like twisted strands of rope winding irregularly over the marial surface. Their appearance has been likened to the crumpled skin on boiled milk. The longest and highest, and the only named example, is the great Serpentine Ridge in Mare Serenitatis. It begins to the east of the ring plain Posidonius and follows a wavy course southwards on the mare giving off several smaller 'tributaries', finally fading out beneath the Promontorium Acherusia at the west end of the Haemus Mountains. East of Posidonius, the ridge bifurcates into two lower ridges at nearly right angles to each other. These, with other lesser ridges northwards, delimit a square area about 50 miles across. The Serpentine Ridge appears to consist of detached arcuate segments like portions of undeveloped elementary rings of considerable height connected by lesser shorter branches. The whole has a folded and corrugated appearance. Neison says that the ridge rises to a height of 700 feet in places.

Fig. 1.14 Wrinkle ridges

A similar but shorter wrinkle ridge stretches southwards from the west side of Plato over the surface of the Mare Imbrium and ends at the much smaller crater Piazzi Smyth. Other notable ridges are found around Timocharis and elsewhere on the Mare Imbrium.

In the Mare Humorum is a group of curved ridges roughly concentric with the west coast. At lunar sunrise they form a fine telescopic object and are reminiscent of ripple marks left on a sandy beach by the receding tide. They extend from an area north of the ring structure Vitello in long wavy lines and gradually peter out towards their southern extremities.

Long ridges straggle over the Oceanus Procellarum around Encke, Marius and Kepler and the area to the north of Aristarchus and Herodotus.

A low ridge runs in a north to south direction in the Mare Crisium and bifurcates to form a diamond-shaped enclosure at its southern end. A more prominent and curved ridge is located in the north–east part of the mare which is concentric with the 'coast'. On the Mare Tranquillitatis is a complex set of ridges in a somewhat radial arrangement around the formation Lamont.

There are concentric ridges on the concave surface of the Mare Nectaris, their appearance suggesting that they may have been formed as a result of the subsidence in stages of the mare floor. However, this does not mean to imply that all ridges are formed in this way. An interesting feature of wrinkle ridges that may have a bearing on their origin is that they are usually found associated with craters of various sizes. Often, small craters sit atop ridges, but more frequently on their sides and near the bases. A prominent crater is usually found at a point where a ridge abruptly changes direction and often other ridges will be found radiating from the crater. Sometimes ridges are cracked along the top and the crack is lighter in color than the rest of the ridge material. It may be that some of the ridges were formed from viscous molten material welling up from cracks in the surface and that others are simply surface manifestations of sudden subsidence of the surface as appears to be the case with the Mare Nectaris ridges and others elsewhere on the Moon's surface.

Domes. Domes are low blister-like surface swellings usually found on the maria or on the dark smooth floors of certain ring formations but also on the bright upland areas. They are found in association with ridges on the maria. The typical dome is circular or roughly circular in outline, often with a pit on its summit and of no great height with very gentle slopes. They can only be distinctly seen at lunar sunrise and rapidly become invisible at higher illumination angles. The average diameter of a dome is about nine miles. They give the impression of being formed by pressure from beneath the surface that was not sufficient to break through completely. A typical example is the isolated dome on the marial surface east of the ring formation Kies in the Mare Nubium. Another isolated dome lies near the small crater Milichius. There is a very large dome on the interior of Darwin and a fine group of them on the floor of Capuanus. Domes generally occur in clusters as in the area around Arago (Mare Tranquillitatis), near Prinz in the Harbinger Mountains, around the crater Hortensius and in other parts of the Oceanus Procellarum. The Lunar Orbiter photography shows that domes are among the commonest features on the Moon's surface.

Clefts (Fig. 1.15). Clefts, or rills as they are sometimes called, are crack-like fissures in the Moon's crust. However, it is unlikely that the majority were formed in the same way as cracks. They are found all over the Moon and vary in length from a few miles to hundreds of miles. They are rarely more

Fig. 1.15 Clefts and valleys

Key

A. Hyginus cleft
B. Ariadaeus cleft
C. Triesnecker clefts
D. Cauchy rilles
E. Sabine and Ritter clefts
F. Goclenius clefts
G. Cleft in Petavius
H. Janssen cleft
I. Sirsalis cleft
J. Hippalus clefts
K. Archimedes clefts
L. Hesiodus cleft

M. Pitatus clefts
N. Alphonsus clefts
O. Posidonius clefts
P. Gassendi clefts
Q. Mersenius clefts
R. Hevelius clefts
S. Parry, Bonpland and Fra Mauro clefts
T. Lacus Mortis clefts
U. Schröter's Valley
V. Rheita 'Valley'
W. Alpine Valley

than a mile wide. Over 2000 are known. Some are so wide as to be visible in small telescopes, such as the great rill on the floor of Petavius, running from the central mountains to the south–east wall. Others are so delicate as to be visible in only the largest telescopes under the best seeing conditions.

Estimates of the depths of clefts vary from about 200 yards to half a mile. The widest are not necessarily the deepest. Narrow ones appear as fine lines of dark shadow at lunar sunrise and may become visible later while the bottoms of the shallower examples can be seen as white lines under higher illumination angles. The direction of a cleft is often continued as a line of hillocks or craterlets.

Sometimes clefts are straight but many are winding and twisting and they may cross over each other. Occasionally they pass right through the walls of craters or through hills.

Some clefts appear to be simple rifts or fractures in the surface, as are those around the formation Letronne. Where there has been either vertical or horizontal movement or both at the sides of a fracture a fault is formed. Conspicuous examples are the faults near the crater Cauchy in Mare Tranquillitatis and the famous 'Straight Wall' in the Mare Nubium. When a strip of ground has fallen between two parallel fractures the resulting trough is called a *graben*. The edges of a graben may be slightly raised. Graben rills or clefts are very common on the Moon.

Clefts often occur in complex systems. To the north–west of the Sinus Medii there is a network of clefts, many of which start a little to the west of the crater Triesnecker. To the north–west of the Triesnecker cleft system is the Hyginus cleft, which looks more like a craterlet chain for most of its length but the Lunar Orbiter frame 96M shows it to be a graben feature, the 'craterlets' actually being collapse pits. One branch extends from Hyginus in a north–east direction and a second branch springs from the opposite side of Hyginus and extends in a nearly westerly direction. This west arm is in a direction parallel to the great Ariadaeus cleft to the west of the Hyginus cleft. These three cleft systems are perhaps the best known examples on the Moon and are easily visible in a three-inch refractor.

Most clefts can be seen only when near the terminator, when they are filled with shadow, but a few, such as the Hyginus cleft, are visible under a high illumination angle and then appear as white lines.

The interiors of many of the large ring structures contain cleft systems, such as that on the floor of Gassendi; here over forty clefts have been recorded. A splendid view of the Gassendi clefts is given in the Lunar Orbiter V frame 178M. There is an interesting group of criss-crossing straight clefts on the floor of Hevelius.

The floor of the great formation Riccioli is traversed by wide shallow graben rills that V.A. Firsoff likens to the marks made by a heavy wheel in soft ground. As seen in Earth–based telescopes, the view is spoiled by foreshortening owing to the nearness of Riccioli to the limb but a very fine 'full face' view of them is given in the Lunar Orbiter IV frame 173,173H. A set of three prominent curved concentric graben rills commences from a point near the crater Vitello on the north 'shore' of Mare Humorum and run towards the crater Hippalus, roughly following the mare 'coastline'. They are well shown in the Lunar Orbiter IV frame 132H.

Some clefts appear to be partly or wholly craterlet chains, such as the previously mentioned Hyginus cleft. The curved cleft near the crater Birt in the Mare Nubium is a craterlet chain throughout its length.

The longest cleft recorded is that associated with the crater Sirsalis in the third quadrant; it is at least 300 miles long.

The so-called sinuous rills meander and wind like river beds. In contrast to other rills whose directions are unaffected by surface relief, the courses of sinuous rills are definitely determined by the surface contours. They usually begin in a crater or an elongated depression and meander from higher to lower ground, becoming shallower or narrower and avoiding all obstacles and higher ground. Examples are the famous Hadley Rill at the base of Mount Hadley in the Apennines, a good view of which is shown in the Apollo frame 15–0587. The Hadley Rill may have been formed as an open lava channel while the surrounding marial lavas were being emplaced. On the other hand, V.A. Firsoff is convinced, after examining every available Orbiter photograph showing sinuous rills, that they are channels worn by flowing water. Similar examples are found near the crater Aristarchus.

The great winding serpentine cleft that commences in Herodotus, known as Schröter's Valley, may be classed as a sinuous rill but it is not typical; there is another meandering rill on its floor. Schröter discovered the serpentine cleft named after him on October 7, 1787. It was the first example of its kind to be discovered. A striking oblique close-up view of it is given in the Apollo frame 15–2611.

Valleys. These are found almost everywhere on the Moon except on the maria. The best known example is the great Alpine Valley previously mentioned, a long straight trough cutting through the Alps. Other considerable valleys are the one south–east of the crater Ukert, the gorge west of Herschel and the wide shallow chasm west of Reichenbach.

The so-called Rheita Valley, named after the crater Rheita at its north end, is situated in the fourth quadrant and is actually a great crater chain. There is a remarkable valley to the south–east of the ring structure Vitruvius. The mountainous southern border of the Mare Crisium is intersected by winding valleys and the ramparts of walled plains such as Ptolemaus are often cut through by valleys.

Bright ray systems (Fig. 1.16). The most extraordinary and enigmatic of the Moon's surface features are the systems of bright rays. They form extensive radiating systems with a crater at or near their apparent point of origin and are brightest and most prominent at full Moon. The rays appear to have no appreciable height above the surface since no visible shadows are cast under even very low illumination angles; they appear to be some kind of surface deposit. The more conspicuous ray systems are those around Tycho, Copernicus, Anaxagoras, Kepler, Olbers, Byrgius A, Aristarchus and Zuchius; lesser systems are those of Aristillus, Autolycus, Timocharis, Proclus, Menelaus, Furnerius A and Theophilus. The most prominent and most extensive ray system is that of Tycho. The Tycho rays are long, thin, straight and bright. They radiate in all directions from Tycho and extend for hundreds of miles, crossing over ridges and other obstructions without being deflected from their course. They resemble lines of longitude radiating from Tycho and give the Moon somewhat the appearance of a peeled orange at the full phase. Among the most eye-catching of the Tychonian rays is one that reaches all the way to Fracastorius and a double one that crosses over the Mare Nubium, passing just to the east of Bullialdus. The white streak running through the middle of Mare Serenitatis is thought by some to originate from Tycho but there is some doubt about this.

Fig. 1.16 Bright ray systems

Key

A. Tycho	*F.* Sirsalis	*K.* Messier and Pickering
B. Copernicus	*G.* Byrgius	*L.* Proclus
C. Aristarchus	*H.* Stevinus	*M.* Thales and Strabo
D. Kepler	*I.* Snellius	*N.* Anaxagoras
E. Olbers	*J.* Langrenus	*O.* Aristillus

W.H. Pickering carefully studied the Tycho rays and discovered the following characteristics:

i The rays do not reach the walls of Tycho but are separated from it by a dark zone which surrounds the crater.

ii The rays do not radiate from the geometrical center of Tycho but originate either from craterlets on the south–east or north rim of the wall or are tangential to the wall.

iii Each ray is not continuous but is made up of a linear array of short plume-like streaks. Generally there is a tiny brilliant crater hardly ever over a mile in diameter at the end of the streak nearer to Tycho and the streak spreads out and becomes fainter at the other end.

Regarding the origin of the Tycho rays, no entirely satisfactory explanation has ever been advanced. Nasmyth and Carpenter likened the rays to the radiating pattern of cracks produced in a glass globe when completely filled with water and heated, the pressure of the expansion of the water breaking the glass. Internal pressure may have cracked the thin crust of the young Moon in a radial pattern and molten material flowing out from the cracks would form a thin surface deposit to give the streaks that we see today. Alternatively, lines of crustal weakness rather than actual cracks may have been formed. Volcanoes at frequent intervals along these lines may have ejected light-colored dust, which blown by expulsion of large volumes of gas from Tycho would give rise to the white streaks with the craterlets at the ends nearer to Tycho.

The ray system of Copernicus is the next most conspicuous and extensive, although individual rays from Furnerius may be longer than the Copernicus rays. The rays of Copernicus spread over marial material and are not as bright as those of Tycho. The Tycho rays themselves are paler where they traverse marial surfaces, so that the difference in brightness is probably not due to an inherent difference in the material forming the rays in the two systems. Unlike the narrow linear rays of Tycho, those of Copernicus are wavy and diffuse forming loops and comet-like plumes. J.E. Spurr likened the appearance of the Copernicus rays to 'snow thinly drifted by a strong wind across a black frozen lake'. Some of the rays do not even trend back into Copernicus but pass right by it. Ridges and other obstructions cause them to change direction, unlike the Tycho rays. Pickering noted that several of the Copernicus streaks start from craterlets within the rim, then flow up the interior and down the exterior wall. Similar craterlets occur within Tycho but streaks emanating from them do not extend far beyond the walls. The major Tycho rays originate outside the 'parent' crater. There is no dark zone outside Copernicus. The rays of Copernicus intersect those of Aristarchus and Kepler and form a complex pattern that is virtually impossible to represent accurately in a drawing. As the solar illumination angle varies the rays seem to vary in position and width.

The rays of Aristillus and Aristarchus are paler than those of Copernicus and Kepler, and those of Theophilus although extensive are very faint and grey in color.

Among the minor ray systems, that of Proclus is interesting. Here, the rays fan out across the Mare Crisium but none is detectable in the opposite direction crossing over the Palus Somnii. As in the case of the Tycho and Copernicus rays, there does not appear to be a single center from which they radiate. There seem to be three sets of rays associated with Proclus, two of them originating from tiny craterlets just outside Proclus and the third from a craterlet on its rim. The rays are often tangential to Proclus, as are those of Tycho, Copernicus, Thales and Anaxagoras. As has been mentioned, the ray systems are brightest at full Moon. As the angle of solar illumination decreases, the rays fade quickly and disappear when it is less than 8°. Their behaviour in this respect is asymmetric because rays on the terminator have been seen in the waning crescent Moon but not in the waxing crescent. Curiously, ray systems at the center of the limb at first or last quarter look as bright as those near the center of the disc at full Moon.

The lunar crater Tycho was photographed by the Orbiters, and the Surveyor 7 space craft landed on its outer slopes and many pictures were transmitted to Earth. Also, chemical analysis of the area was performed. Even with this information and the fact that astronauts have actually walked on the lunar surface in areas covered by ray material, we still do not understand what the Moon's bright rays really are or what process or processes gave rise to them.

Further Reading

Books–General

New Guide to the Moon. Moore, P.A. W.W. Norton and Co., Inc., United States of America (1976).

Structure of the Moon's Surface. Fielder, G. Pergamon Press, United States of America (1961).

Strange World of the Moon. Firsoff, V.A. Hutchinson, London, England (1959).

The Old Moon and the New. Firsoff, V.A. Sidgwick and Jackson, London, England (1969).

Origin of lunar surface features

i *Impact hypothesis*

The Face of the Moon. Baldwin, R.B. University of Chicago Press, Chicago, Illinois, United States of America (1949).

The Measure of the Moon. Baldwin, R.B. University of Chicago Press Chicago, Illinois, United States of America (1963).

ii *Igneous hypothesis*

Surface of the Moon. Firsoff, V.A. Hutchinson, London, England (1961).

Lunar Geology. Fielder, G. Lutterworth Press (1965).

The Craters of the Moon. Moore, P.A. and Cattermole, P. Lutterworth Press (1967).

Geology Applied to Selenology. Spurr, J.E. Vol. 1 (1944); Vol. 2 (1945); Vol. 3 (1948); Vol. 4 (1949). Lancaster, Pennsylvania, United States of America.

Books based on space craft data and close-up photography

Lunar Panorama. Lowman, P.D. Weltflugbild. Reinhold A. Müller. Feldmeilen/Zurich, Switzerland (1969).

Geology of the Moon. A Stratigraphic View. Mutch, T.A. Princeton University Press (1970).

The Moon as Viewed by Lunar Orbiter. Kosofsky, L.J. and El-Baz, F. National Aeronautical and Space Administration, Washington, DC (1970).

Moon Morphology. Schultz, P.H. University of Texas Press (1972).

Lunar Science: A Post-Apollo View. Taylor, S.R. Pergamon Press, Inc. (1975).

Geology on the Moon. Guest, J.E. and Greeley, R. Wykeham Publications Ltd, London, England (1977).

2

The Moon's Motions and Consequent Phenomena

The branch of astronomy that is concerned with the motions of the Moon is known as Lunar Theory. It involves extremely complex mathematics, dealing as it does with the many different perturbations of the Moon's orbit, orbital motion and axial rotation caused by the gravitational pulls of the Earth, Sun and other bodies in the Solar System. It has taxed some of the best mathematical minds. In this chapter, a simplified description of the Moon's motions will be given. Of necessity, much will be omitted and only those phenomena that are of direct and practical interest to the lunar observer will be included.

The Celestial Sphere

When we look up at the night sky the stars and other celestial bodies appear to be attached to the inner surface of a great hemispherical bowl, and we ourselves seem to stand at the center of a more or less plane surface extending to the horizon with the great inverted bowl of the sky above us. The other half of the bowl is out of sight beneath our feet and is continuous with the hemisphere above. This great heavenly globe is called the Celestial Sphere (Fig. 2.1).

If we watch the night sky hour by hour, the stars slowly drift in an east to west direction while retaining the same positions relative to one another in the well-known patterns called constellations. It is as if the whole Celestial Sphere is slowly rotating.

Actually, of course, there is really no Celestial Sphere and the stars and other heavenly bodies are all at various immense distances from us, as we learned in the last chapter. Their appearance of being all at the same distance and projected onto the inner surface of a sphere like the star images in a planetarium is an illusion.

Although ficticious, the concept of the Celestial Sphere is useful for describing the motions and positions of the the stars, planets, Sun, Moon and other sky objects. In reality, it is our own Earth rotating in a west to east direction about its axis that gives the appearance of the stars rising above the eastern horizon, drifting in an east to west direction and then setting at the western horizon.

The ends of the Earth's axis are at the north and south poles, so that if the axis is prolonged at both ends the points where they intersect the Celestial Sphere are the north and south Celestial Poles. It is around this extension of the Earth's axis that the Celestial Sphere appears to rotate.

The point on the Celestial Sphere directly overhead is called the zenith. The equator is the great circle passing around the Earth, the plane of which is perpendicular to the axis and divides the Earth into northern and southern hemispheres. Its projection onto the Celestial Sphere is the Celestial Equator and it cuts the horizon exactly at the east and west points. Just as there are smaller imaginary circles of latitude on the Earthly sphere parallel to the equator representing angular distance north or south from the equator, so their projections, the circles of celestial latitude onto the Celestial Sphere,

Fig. 2.1 The Celestial Sphere

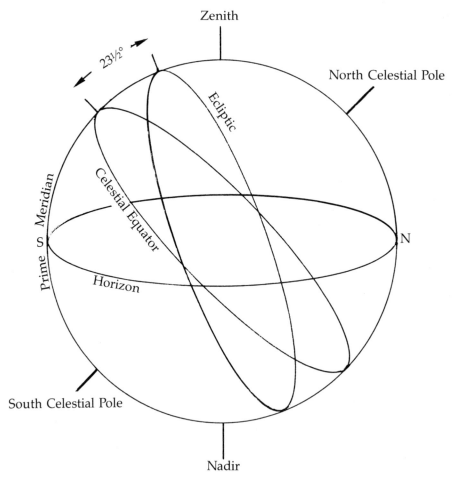

represent angular distance, or *Declination* as it is called, north or south of the Celestial Equator. Similarly, the meridians of longitude on the Earth's surface, imaginary great circles passing through and intersecting at the poles and perpendicular to the equator, when projected onto the celestial sphere from the meridians of celestial longitude, or *Right Ascension* as it is called. Each of these circles is called an hour circle and there are 24 of them on the Celestial Sphere at intervals of 15°. Right Ascension is customarily measured in hours, minutes and seconds instead of degrees because any one of the hour circles makes a complete circuit of the heavens in about 24 hours. The Right Ascension of a star is measured eastward all the way around the Celestial Equator from a fixed zero hour circle to any other hour circle or intermediate position. The position of a celestial body on the Celestial Sphere is therefore defined by its Right Ascension and Declination, just as a place on the Earth's surface is defined by its longitude (degrees east or west of the Greenwich Meridian) and latitude respectively.

The imaginary great circle intersecting the Celestial Equator at right angles and passing through the observer's zenith and the celestial poles is the Prime Meridian on the Celestial Sphere for that observer. It intersects the horizon at exactly the north and south points. When any celestial body passes across this meridian it is said to transit or to culminate. At this moment the object is at its highest point in the sky on that particular day. Before culmination the object is 'rising' and afterwards it is 'setting'.

The axis of the Earth is tilted, so that the equator is inclined at an angle of 23½° to the plane of the Earth's orbit around the Sun. During the course of one year of 365¼ days, the Earth makes one complete journey around the sun. During this time the Sun appears to drift slowly in a west to east direction against the background of fixed stars at the rate of slightly less than 1° per day. This annual apparent path of the Sun traces out a great circle on the Celestial Sphere called the Ecliptic that is tilted at an angle of 23½° to the Celestial Equator. It therefore intersects the plane of the Celestial Equator at two points. These are called Equinoxes. Because of the gravitational pull of the sun and the moon, Equinoxes drift slowly aound the Ecliptic in a westerly direction, one complete circuit taking 26,000 years. This movement is called the Precession of the Equinoxes. On or about March 21 of each year, the Sun is at the Spring or Vernal Equinox, the point at which it passes from the south to the north side of the Ecliptic. At the Summer Solstice, around June 21, the Sun is at that point of the northern half of the Ecliptic midway between the two Equinoxes and is at its maximum Declination of 23½° north of the Celestial Equator. It is therefore at its highest point in the sky at a given latitude on the Earth when it transits the Prime Meridian on this date. Its actual angular height above the horizon will depend on the observer's Earthly latitude.

Around September 21 the Sun is at the Autumn Equinox and passes from the northern to the southern half of the Ecliptic, reaching its maximum declination south of the Celestial Equator on the Winter Solstice on or around December 21 when it is again midway between the two Equinoxes and diametrically opposite the Summer Solstice. It is at its lowest point on the observer's Prime Meridian when it transits on this date and its declination is 23½° south of the Celestial Equator. Again, the actual angular height above the horizon depends on the observer's latitude. The Sun's angular altitude when on the Prime Meridian therefore varies by 47° during the course of a year.

Now that the Celestial Sphere and the various circles upon it have been described we may go on to consider the Moon's motions and the phenomena that these give rise to. Matters of considerable practical interest to the Lunar observer are the apparent changes of shape or 'phases' exhibited by the Moon, the times of the day and month when the Moon is visible and the best times of the day, month and year for observing the Moon (apart from the effects of weather and state of the atmosphere).

The phases of the Moon

Everyone is familiar with the Moon's 'phases', the continuous changes in apparent shape of the Moon, one complete cycle of which takes slightly less than a calendar month. The very slender crescent Moon seen just after the 'new Moon' in the western sky at sunset gradually thickens on each successive evening and at the same time the Moon moves further away from the Sun in an easterly direction. About seven days after new Moon the crescent has expanded into a semicircular shape ('half Moon'), the phase called first quarter. The Moon is due south and on the meridian at about sunset. On the following evenings, the straight edge bulges more and more into a convex shape giving the phase called 'gibbous', until about a week after first quarter the Moon appears as a complete circular disc of light, the 'full Moon'. It is now diametrically opposite the Sun on the Celestial Sphere and rises in the eastern part of the sky at about sunset. During this part of the cycle when the Moon appears to be increasing in size and brightness, it is said to be 'waxing'.

The second half of the phase cycle repeats the first half in reverse order. The Moon continues to rise later and later each night, the edge of the disc opposite from the one that has been changing shape becomes less and less convex and the gibbous shape is assumed again. About a week after full Moon we have a half Moon ('last quarter') the opposite way around from first quarter and the Moon transits the meridian at sunrise. (Perhaps it should be explained here that the half Moon phases are called 'quarters' because at first quarter the Moon has completed the first quarter of its cycle of phases and at last quarter it is about to commence the last quarter of the cycle.) If we care to stay up into the early hours of the morning, the last quarter Moon will be seen to change into a thin crescent shape that becomes thinner on each successive night. When very thin, it is lost in the light of the rising Sun at dawn. During this second half of the phase cycle when the Moon is apparently decreasing in size it is said to be 'waning'. For a few days the Moon is lost from sight as it is too close to the Sun and it is then said to be at the 'new' phase. When we next see the familiar thin crescent in the western sky after sunset, the Moon may well seem to say to us 'since you last saw me at this phase, nearly a whole month has gone by'. The complete cycle of phase changes from one new Moon to the next is called a *lunation*.

The varying phases of the Moon are a consequence of the facts that the Moon is not self-luminous, shining only by light reflected from the Sun, and that it journeys around the Earth in a nearly circular orbit. The phases are simply the different portions of the Moon's illuminated hemisphere that we see as a result of the varying positions of the Moon relative to the Earth and Sun during its orbital motion around the Earth (Fig. 2.2). For example, at position *A*, a thin sliver of the illuminated hemisphere is visible, which we see as a crescent Moon, whereas at other positions we see varying amounts of the illuminated hemisphere at the different phases. The line connecting the tips of the crescent Moon is always at right angles to the great circle on the Celestial Sphere upon which the Sun and Moon are both situated at the time.

The Moon's phases can easily be simulated by taking a tennis ball held at arm's length towards a bright lamp. There should be no other light in

Fig. 2.2 The cause of the Moon's phases (not to scale)

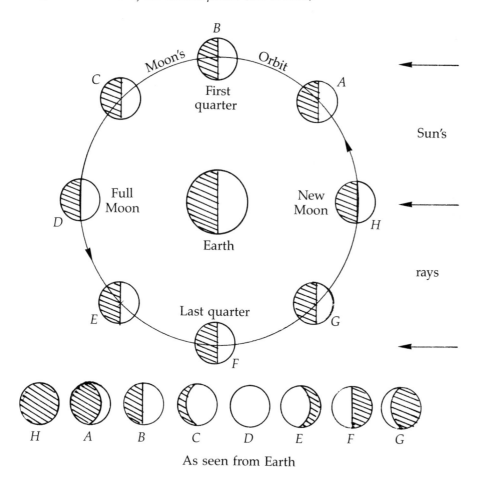

the room. The lamp represents the Sun, the tennis ball the Moon and our head the Earth. Now, slowly turn around from right to left still holding the tennis ball directly in front of you. As you turn you will see more and more of the illuminated half of the tennis ball, which simulates the waxing phases of the Moon. After 'full Moon', when your head is between the lamp and the tennis ball, you will see the waning phases and after one complete circuit the dark hemisphere of the tennis ball – 'new Moon' – will be seen.

The bright permanent circular outline of the Moon is called the *limb* and the boundary between the light and dark hemispheres that appears to change shape and passes across the Moon during a lunation is called the *terminator*. The terminator that we see during the waxing cycle marks the position of sunrise on the Moon's surface and it is therefore called the sunrise or morning terminator. Likewise, on the terminator of the waning Moon the Sun is setting and so it is called the sunset or evening terminator (Fig. 2.3).

Ignorance of the cause of the Moon's phases has sometimes led to errors in art where the Moon is represented in pictures or in literature. For instance, the crescent Moon has been shown in art at times with the horns of the crescent pointing towards the Sun, clearly an impossibility as will now be realised. Then, in Coleridge's *Rime of the Ancient Mariner*, there occur these lines:

> Till clombe above the eastern bar
> the horned moon with one bright star
> within the nether tip

Again, it is clearly impossible for a star to be seen between the tips of the lunar crescent because the unilluminated solid body of the Moon would be in that position and would conceal any star, even quite a way from the crescent.

The orbit of the Moon is inclined at an angle of 5° 8′ 43″ to the plane of the Ecliptic and the two points where it intersects the Ecliptic are called *nodes*. One consequence of this tilt is that our terrestrial view of some of the Moon's phases is slightly skewed; the observer's geographical location also contributes to this effect. The result is that we sometimes 'underlook'

Fig. 2.3 Limb and terminator

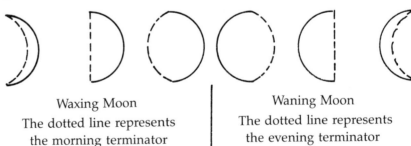

Waxing Moon	Waning Moon
The dotted line represents the morning terminator	The dotted line represents the evening terminator

and sometimes 'overlook' the Moon by a slight angle. At the crescent phase, this means that we will see a little more at one cusp and a little less at the other, a phenomenon called the *polar phase* or *phase defect*. This is especially noticeable at the new Moon, when a thin line of light will be seen at one or other pole, this depending on whether the Moon is north or south of the Ecliptic and on its apparent height in the sky. At full, a slight shadow will be seen at the other pole. These effects are not apparent at the half Moon phases (dichotomy). A true new Moon can be seen only at a solar eclipse and a true full Moon during a lunar eclipse (see below).

The 'Earthshine'

A further consideration of Fig. 2.2 will show that an imaginary lunar inhabitant will see the Earth go through a similar cycle of phases in the course of about one month. At full Moon our Lunarian will see a 'new' Earth and a 'full' Earth will be seen from the Moon's dark side. All of the other intermediate phases will also be seen and will be the exact reverse of the Moon's appearance from the Earth.

A really bright full Moon sheds a lot of light on the Earth at night and this is especially so in the tropics, where a full Moon can be bright enough to read by. It follows that a 'full' Earth will similarly light up the new Moon's dark side. Since the Earth is so much larger and more reflective than the Moon owing to its cloudy atmosphere, the amount of light it reflects on to the Moon's dark side will be much greater. We can see this effect vividly on a clear dark evening when the Moon is a thin crescent just after new. The dark part unilluminated by direct sunlight can often be distinctly seen because of the light reflected by the Earth, which will be nearly full as seen from the Moon. The effect is often charmingly referred to as 'the old Moon in the new Moon's arms'.

The Sidereal and Synodic months

The Moon makes a complete orbit of the Earth in approximately one month; in fact the word 'month' is derived from 'Moon'. If we determine how long it takes for the Moon to return to the same fixed point in its orbit after one complete journey around the earth, as indicated by its return to the same point among the background of the stars, we will find that it takes about 27⅓ days. This is the true orbital revolution period and is called the Sidereal month. However, if we determine the time taken from a given phase to exactly the same phase in the next lunation, say from new Moon to new Moon, the time taken is about 29½ days. This is called the Synodic month. The explanation for the difference in the Sidereal and Synodic months is shown in Fig. 2.4. At position *A*, the Moon is 'new'. After a complete revolution in its orbit it is at position *A* again but in the meantime the Earth has moved on in its orbit around the Sun and so the Moon has to move further past point *A* to point *B* before it is 'new'. This extra time of more than two days needed to arrive at the new phase is the cause of the difference between the Sidereal and Synodic months.

Fig. 2.4 The Sidereal and Synodic months (not to scale)

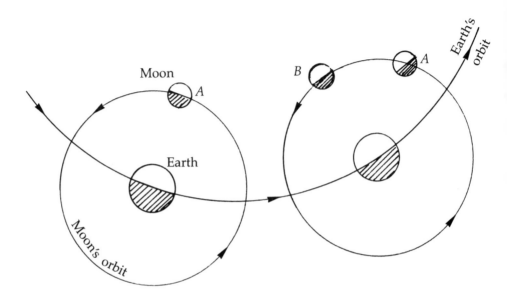

The Moon's path in the sky

Even a casual observer will notice that the angular height of the Moon above the horizon when at its highest point in the sky varies considerably at different phases during a lunation and that its height at its highest point at given phases varies through the year. The positions on the horizon of rising and setting during the lunation and for a given phase during the year also show wide variation. These changes in the Moon's apparent path in the sky are more dramatic and rapid than those of the Sun because whereas the Sun takes a year to complete one circuit of the Celestial Sphere, the Moon takes slightly less than a calendar month.

As previously noted, the Moon's orbit is inclined at an angle of about 5° to the plane of the Ecliptic and the points of intersection of the Moon's and the Earth's orbits are called nodes. The line of joining them is called the *line of nodes*. The Moon passes from south to north of the Ecliptic at one node and from north to south at the other. They are therefore referred to as the *ascending* and *descending* nodes respectively. If the plane of the Moon's orbit coincided with that of the Earth's, its angular meridian height above the horizon, like the Sun's, would vary by 47° during the year. However, because of the 5° orbital tilt to the Ecliptic, it can pass 5° higher than the

Sun's most northerly declination above the Celestial Equator and so can reach a maximum northerly declination of 28½°. Likewise, it can pass 5° lower than the Sun's maximum southerly declination so that the Moon can attain a maximum southerly declination of 28½°. Hence, the meridian altitude of the Moon can vary by as much as 57° during the course of a year.

We will now trace the path of the Moon in the sky during the lunation at different times of the year. For simplicity, assume that the Moon's orbit is in the same plane as the Earth's so that the Moon's path in the sky will coincide with the Ecliptic. On or around March 21, the Sun is at the Spring Equinox and sets at or very close to due west. The northern half of the Ecliptic is therefore above the horizon at sunset at this time. The first quarter Moon would be on the meridian in the same place that the Sun would be at Summer Solstice and so rides high in the spring skies. It sets north of the west point of the horizon. Between first quarter and full Moon, the meridian height of the Moon gradually decreases until at full, when the Moon rises due east at sunset, it is at the Autumnal Equinox and its meridian height is much lower when it transits at midnight as it is now on the Celestial Equator. It sets at the west point of the horizon. The last quarter Moon will be at the Winter Solstice. It rises a little south of east at midnight and is low down in the south when it transits. It sets around noon, south of the west point.

Three months later at the Summer Solstice, the Sun sets in the north-west and the first quarter Moon is now at the Autumnal Equinox and follows the Celestial Equator, so that its summer meridian height is lower than it was in the spring. It rises due east and sets due west. The full Moon, now at the Winter Solstice, rises well south of east, transits low in the south at midnight and sets well south of west. It does not spend many hours in the sky at this time of year. The last quarter Moon at the Spring Equinox, now appears higher in the sky than in spring when it crosses the meridian.

At the beginning of autumn, on or about September 21, the Sun's setting point has shifted back due west and at sunset the whole of the south part of the Ecliptic is above the horizon. The first quarter Moon occupies the full Moon's position at the Winter Solstice and its meridian height is lowest at this time of year. The full Moon, at the Vernal Equinox, is on the equator and is higher when on the meridian than in summer. The last quarter Moon, now at the Summer Solstice, is at its greatest meridian height in the cool mornings of autumn.

Three months later at the beginning of winter around December 21, the first quarter Moon at sunset will have climbed higher in the sky, as it is now located at the Vernal Equinox. It rises due east and sets due west, following the Celestial Equator in its path across the sky. The full Moon, now at the Summer Solstice, rises north of east and is higher in the sky, as it transits the meridian, than at any other time of the year. It sets north of west. The last quarter Moon is at the Vernal Equinox and rises due east. Its meridian height has decreased from what it was three months ago and it sets due west.

Actually, the Moon's path in the sky does not coincide exactly with the Ecliptic and is not fixed, owing to the angular tilt of its orbit to the Ecliptic of about 5°. Further, because of the gravitational pull of the Sun, the nodes slowly shift their position in a westward direction, one complete revolution around the Ecliptic taking 18.61 years. This movement is called the *Regression of the Nodes*. It is analagous to the precession of the Equinoxes but, of course, is much faster. Suppose that the ascending node is at the Autumnal Equinox. At new Moon, around the beginning of spring when the Sun is at the Vernal Equinox, the Moon's path during the first half of the lunation will be south of the Ecliptic and its greatest distance north of the Equator will be 18½° (23½°−5°). During the second half of the lunation the Moon will be north of the Ecliptic and its maximum distance south of the equator will be 18½° i.e., 5° north of the Ecliptic. The Moon's path will therefore be entirely within the area of sky bounded by the Equator and Ecliptic (Fig 2.5a).

After four years and eight months, i.e., one quarter of a revolution of the nodes around the Ecliptic, the ascending node will be at the Summer Solstice. The Moon's path now lies half outside and half inside the area bounded by the Equator and Ecliptic (Fig 2.5b).

After another four years and eight months, i.e., 9.3 years after the time when the ascending node was at the Autumnal Equinox, the ascending node will be at the Vernal Equinox. The Moon's path will now lie entirely outside the area bounded by the Ecliptic and Equator (Fig 2.5c).

After the lapse of another four years and eight months, the ascending node will be at the Winter Solstice and the Moon's path will again be half within and half without the area bounded by the Ecliptic and Equator (Fig 2.5d). Finally, after a further four years and eight months the ascending node will be at the Autumnal Equinox again, i.e., 18.61 years after it was last there.

From the foregoing it follows that the Moon will always be found within a band of sky 10° wide bounded by two great circles parallel to the Ecliptic, one 5° north and the other 5° south of it.

The 'Harvest Moon'

During the course of 24 hours the Moon advances eastward in its apparent path against the background of fixed stars by about 13°. If the Moon moved about the Celestial Equator it would rise later each night or day by an almost equal time interval on every day of the year and its path in the sky would make the same angle with the horizon at all times, the angle depending on the observer's latitude. Also, it would rise and set due east and west throughout the year. However, since the Moon actually moves close to the Ecliptic, the angle that its path makes with the horizon varies considerably at different times of the year and this causes wide variations in the intervals of time, or *retardations* as they are called, between successive Moon risings during the year. The mean value of the retardation is about 50½ minutes. Also, the rising and setting points on the horizon shift their positions considerably during the year. The same is also true of the Sun but, since the Moon makes one complete circuit of the heavens much more quickly than the Sun, the different intervals between rising times, the angle that the Moon's path makes with the horizon and the shift in the positions of the rising and setting points are much more perceptible than the Sun's. Fig 2.6 shows the situation at Moonrise at the Spring and Autumn Equinoxes in the Northern Hemisphere. In spring on March 21 the Sun sets due west and the full Moon rises diametrically opposite at the east point of the horizon at or near the position occupied by the Autumnal Equinox. At this

Fig. 2.5 The Moon's apparent path in the sky

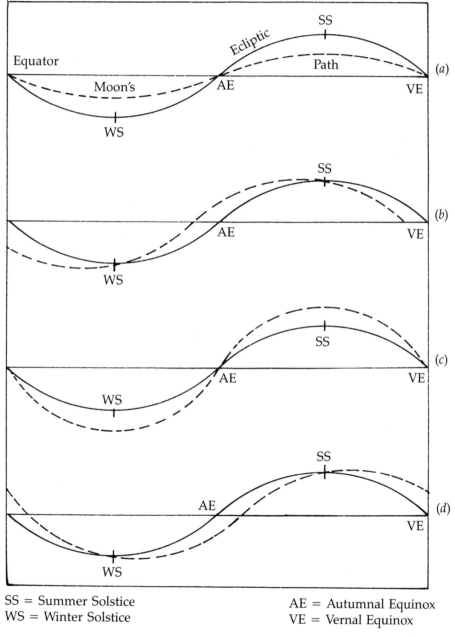

SS = Summer Solstice   AE = Autumnal Equinox
WS = Winter Solstice   VE = Vernal Equinox

moment the whole of the northern half of the Ecliptic is above the horizon and makes its maximum angle with it. Twenty-four hours later the Moon will have moved 13° eastward to position 2 and so has to climb the vertical distance *h* to reach the Moonrise position. Under these conditions the time between successive risings of the Moon is maximal and is about 1 hour and 20 minutes (at about the latitude of London); the actual value will depend on whether the Moon is exactly on the Ecliptic or north or south of it by the amount determined by the inclination of its orbit. In autumn on September 21 at sunset the whole of the southern part of the Ecliptic is above the horizon and the angle made with it is much shallower than in the spring. Hence, although the Moon still travels eastward by 13° in 24 hours the vertical distance *h* that it has to climb from position 2 to the rising position is now much less. At this time of year the retardation can be as little as 15 minutes and when about full the Moon appears to rise at nearly the same time for several successive nights. The full Moon of September is therefore called the Harvest Moon because at one time farmers found the extra source of light valuable at this time of year when they were gathering in the harvest. A similar effect is noticed in October, the full Moon of this month being called the 'Hunter's Moon', but the retardation is greater at this time.

Another factor that effects the retardation in addition to the inclination of the Moon's orbit to the Ecliptic is the Moon's variable motion caused by the eccentricity of its orbit (see next section). The combined effect of these two factors is that sometimes the Harvest Moon will rise only 9 or 10 minutes later each night, and at other times at the same season, the intervals between successive risings may be more than 30 minutes. Similarly in spring, the actual intervals between successive full Moon risings may be from about 70 minutes to as much as 90 minutes in different years.

At latitudes further north than London, the phenomena of the Harvest and Hunter's Moon are more pronounced.

Fig. 2.6 The 'Harvest Moon'

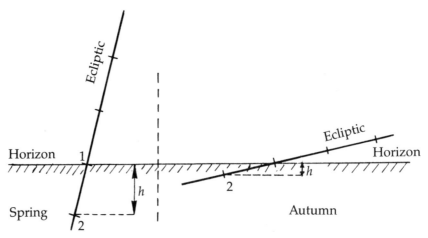

Form of the Moon's orbit

So far it has been assumed for simplicity that the Moon's orbit around the Earth is circular. Actually, it is only approximately circular and is really an ellipse, as are also the orbits of the planets around the Sun and the orbits of satellites around their planets.

An ellipse is a closed curve sometimes called an 'oval' and can be drawn as follows. Place a piece of white paper on a smooth wood drawing board and firmly press two drawing pins or thumb tacks about three inches apart through the paper into the board. Leave a little space between the heads of the pins and the paper so that a loop of strong twine or thread may be slipped over them. Draw the thread tight over the paper with the point of a pencil and move the pencil around the pins holding the thread taut all the time as you move the pencil. In this way you will draw a closed curve - the ellipse (Fig. 2.7).

This method of drawing an ellipse is based on the definition of an ellipse, which is the locus (path) of points in a plane, the sum of whose distances from two other fixed coplanar points (represented by the pins) is constant. The two fixed points are called the *foci* (singular = *focus*) of the ellipse. For a given length of thread the further apart are the pins (foci) the more elongated will be the ellipse and the closer they are together the more will the ellipse approximate to a circle. A circle may in fact be regarded as a limiting form of the ellipse in which the foci are coincident.

A straight line drawn through the centre of a circle from one side to the other defines the diameter of the circle and it is the same length in all directions. An ellipse on the other hand possesses a major axis passing

Fig. 2.7 Drawing an ellipse

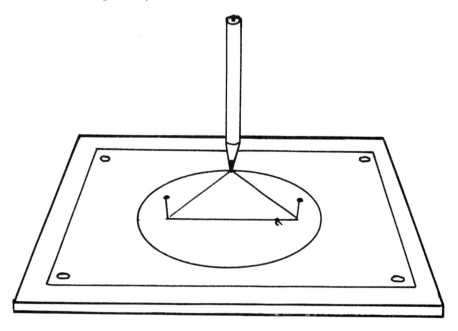

through the foci and a minor axis perpendicular to this and passing through the center of the ellipse (Fig. 2.8).

The degree of deviation of an ellipse from a circular form is measured by its *eccentricity*, which is defined as the ratio:

$$\frac{\text{distance along major axis from centre to focus}}{\text{distance along major axis from centre to boundary of ellipse}}$$

Thus an ellipse that differs only slightly in shape from a circle has a small eccentricity and an elongated ellipse has a large eccentricity. The orbits of the planets around the Sun and of satellites around planets including our Moon all have very small eccentricities and approximate closely to circles. The orbit of our Moon has an average eccentricity of about 0.0549. The value varies slightly owing to the distorting effect of the gravitational pull of the Sun. The Earth's centre occupies one of the foci of the Moon's elliptical orbit and so the Moon's distance from the Earth varies during its orbital revolution in the course of a month. When it is at its nearest to the Earth (about 221 460 miles) it is said to be at *perigee* and at its farthest (about 252 700 miles) it is at *apogee*. Similarly the Sun's center occupies one focus of the Earth's elliptical orbit and the Earth at its greatest distance (94 600 000 miles) and least distance (91 400 000 miles) from the Sun *aphelion* and *perihelion* respectively.

The line joining the perigee and apogee is called the *line of apsides* Fig. 2.9). Because of the varying distance during the lunation the Moon's apparent angular size as seen from Earth also varies from a maximum of 33′ 30″ at perigee to a minimum of 29′ 26″ at apogee (Fig. 2.10). There is another

Fig. 2.8 Major and minor axes of an ellipse

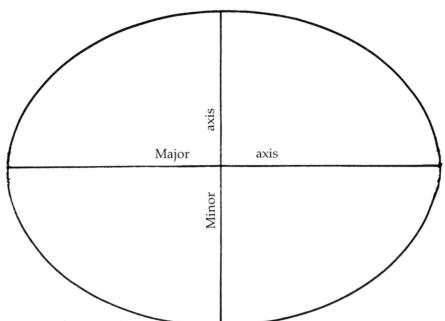

factor that affects the apparent angular size of the Moon. Its distance from us is about 60 Earth radii. Between the time that the Moon rises and transits the meridian, the Earth's rotation has brought the observer nearer to the Moon by a distance equal, at most, to an Earth radius. The Moon's distance from the observer has therefore been lessened by 1/60 of what it was at moonrise and the Moon's apparent angular diameter has increased by about 30″. However, this is not noticeable to the eye although it is detectable with instruments.

Fig. 2.9 The line of apsides (not to scale)

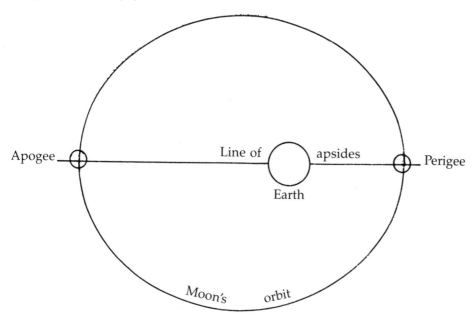

Fig. 2.10 Variation in apparent size of the Moon

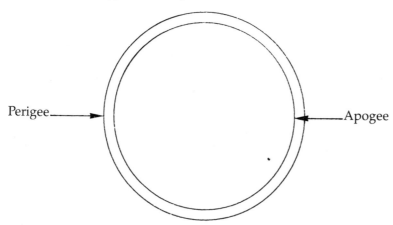

The Moon's orbital motion, axial rotation and librations

As we have already noted, the Moon makes one complete circuit of its orbit around the earth in the Sidereal month of about 27⅓ days. During this same time the Moon makes one complete rotation on its axis and so has what is called *synchronous* or *captured* rotation, a result of the Earth's gravitational pull. The Moon therefore always keeps practically the same hemisphere turned towards the Earth.

In accordance with Kepler's Second Law of Planetary Motion the Moon's orbital speed is not uniform: it is fastest at perigee and slowest at apogee, the radius vector, i.e., the line joining the orbital focus occupied by the Earth and the center of the Moon, sweeping out equal areas in equal times (Fig. 2.11). However, the Moon's axial rotation speed is constant and so at some positions in its orbit the Moon's rotation will lag behind its orbital speed at other positions it will exceed it. Therefore, as seen from the Earth, the Moon will appear to undergo a slow wobbling motion from side to side about a vertical axis. At certain times we are thus able to see a little further around the east limb and at others a little further around the west limb (lunar cardinal points are explained in the next Chapter). This apparent wobbling is called *libration* and in this case is libration in longitude. It is called libration because of the resemblance to the slow see-sawing motion

Fig. 2.11 Equal areas in equal times (Kepler's Second Law of Planetary Motion)

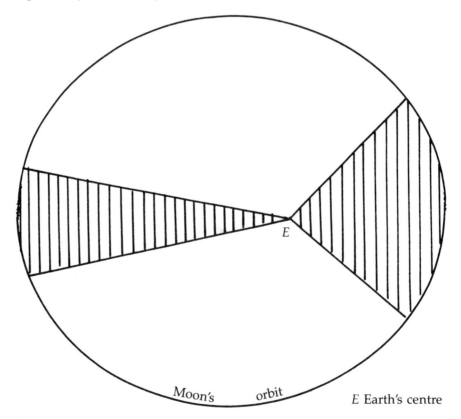

Moon's   orbit   E Earth's centre

of scales about a fulcrum when they are slightly off equilibrium. In fact both the name of the Zodiacal constellation Libra (the Scales) and the word equilibrium come from the same Latin root meaning 'balance'. The longitudinal libration can attain a maximum value of about 7° 57' on either side of the mean center of the Moon's disc.

The *diurnal libration* is an effect due to parallax and is caused by the shifting viewpoint of the observer as a result of the Earth's rotation. It is mostly longitudinal and we can see a little more around the Moon's western limb when it is rising and a little more around the eastern limb when it is setting. This libration amounts to ± 57' 2.6".

We have already noted that the Moon's orbit is inclined at an angle of about 5° to the plane of the Ecliptic. The axis of rotation is inclined to the plane of the Moon's orbit at an angle of 1° 30'. Both of these can therefore combine to give a libration in latitude with a maximum tilting of either pole towards or away from the observer of about 6° 51'.

The position of the observer on the Earth's surface gives rise to a latitudinal parallactic effect which permits someone in the northern Hemisphere to see a little beyond the Moon's northern limb and in the Earth's southern Hemisphere a little further beyond the Moon's southern limb; as with longitudinal diurnal libration this amounts to ± 57' 2.6". The combined effect of these two librations is to give a maximum libration in latitude of ± 7° 48' 2.6".

Because of the longitudinal, latitudinal and parallactic librations the observer is able to see over an extended period of time a little more than 50 per cent of the Moon's surface; about 59 per cent of the total surface is thus made visible, the remaining 41 per cent being forever hidden from Earth-bound observers. However, this little extra of the Moon's surface brought into view by the libration is not seen to advantage because of the foreshortening due to its closeness to the limb. These regions around the lunar limb that are alternately visible and invisible owing to the librations amount to about 18 per cent of the Moon's total surface area. The effects of libration are especially noticeable at the north-east, south-east, south-west and north-west limbs. Data on which limb regions of the Moon are favorably exposed throughout the year are published in the *Handbook of the British Astronomical Association* and in the *Observer's Handbook* of the Royal Astronomical Society of Canada.

In addition to these optical librations as they are called there is also a very small but real physical libration due to the fact that the Moon is not quite a perfect sphere. There is a slight bulge extending towards the Earth and the factors that give rise to the longitudinal and latitudinal librations cause the Earth's gravitational pull on this bulge to change direction slightly. This produces a slight irregularity in the Moon's axial rotation, the *physical libration*. Referred to the Moon's center, the amplitude of this never exceeds 2', which amounts to 0.54" at the center of the disc as viewed from the Earth. In practice, this is usually below the limit of telescopic resolving power and so is of no practical importance to lunar observers.

As a result of the combined effects of the optical librations any point on the Moon's surface will appear to oscillate and slowly trace out a connected series of elliptical paths which continually vary in different lunations with respect to position, size and the directions in which they are described. On average, a complete cycle of libratory changes takes about six years. During this cycle there are transient close passes to the starting position but they are approached and left in different directions. This six-year cycle is not related to the Moon's phases, so that when a given point returns to its original libratory condition after the six-year cycle it will be seen under quite different illumination conditions from when it commenced the cycle. It may even be on the unilluminated hemisphere. Now, both libration and solar illumination angle greatly affect the appearance of lunar surface features. It can be shown that because of the interplay of these two combined effects a given lunar surface feature seen under a given combination of illumination and libration conditions will not be seen under precisely similar conditions again until a period of 186 years has elapsed. Therefore a lunar feature studied by a given observer will never be seen under exactly identical conditions again in his lifetime. However, two or maybe three times in a decade a given feature of the Moon will be seen under conditions so nearly alike that its aspect will be sufficiently similar so that, providing weather and seeing conditions are favorable, valid comparisons with previous observations may be made with respect to possible changes in physical form of surface features. Nevertheless, such opportunities are always going to be quite rare during the lifetime of any one observer who wishes to make reliable comparative studies of any part of the Moon's surface. The role of solar illumination angle and libration in the appearance of lunar features and the opportunities to observe given features more than once under nearly identical combinations of both are critical factors in testing the validity of alleged 'changes' in, or 'appearances' and 'disappearances' of, lunar objects. This will be referred to again in Chapter 6.

Solar and lunar eclipses

If the plane of the Moon's orbit coincided with that of the Ecliptic then at every new Moon the Earth, Moon and Sun would lie in a straight line and the Moon would pass in front of the Sun as seen from the Earth. By a curious coincidence, the apparent angular diameter of the Sun and Moon are very closely similar. The Sun is very much larger than the Moon but is much further from us and at such a distance that it looks almost exactly the same size as the Moon. Depending on the exact geographical location of the observer, the Moon when new will appear to pass either exactly in front of the Sun giving a total eclipse or off center so as to only partly cover the Sun giving a partial eclipse. The different views are of course due to the parallactic effect. The reason why a solar eclipse does not occur at every new Moon is another result of the Moon's orbit being inclined to the plane of the Ecliptic. The tilt is sufficient to carry the Moon clear of the Sun at new Moon, so that an eclipse is not therefore seen every time there is a new Moon. However, if the Moon is at or very near to a node at the time of new Moon, then the Earth, Moon and Sun will be in a nearly straight line and a solar eclipse will be the result. The shadow cast by the Moon consists of two parts, a long tapering cone of full shadow called the umbra, within which no part of the Sun can be seen, and an expanding cone of partial shadow called the penumbra, from within which part of the Sun can be seen (fig. 2.12). This shadow structure is a result of the Sun not being a point source of light.

If when the conditions for a solar eclipse are met and the Sun is at aphelion and the Moon at perigee, then the tip of the umbra of the Moon's shadow will touch the Earth and a total solar eclipse will be seen within the area covered by the umbral shadow. Since the Earth is at its greatest distance from the Sun and the Moon at its least distance from the Earth, as stated above, the apparent angular size of the Moon's disc is just sufficient to cover the Sun's disc completely. A partial eclipse will be seen in the areas of the Earth within the penumbral shadow.

If the distance conditions are the reverse with the Sun at perihelion and the Moon at apogee, the tip of the Moon's umbra will fall short of the Earth's surface; the Moon will appear to be not quite large enough to cover the Sun's disc. At those parts of the Earth where the central passage of the Moon across the Sun is visible, a thin ring of the Sun will be seen around the Moon's circumference giving an *annular* eclipse. The annular ring is of maximal width under these conditions. Annular eclipses may still be seen if the Sun and Moon are not quite at their minimal and maximal distances respectively.

An eclipse of the Moon occurs when the Moon passes into the shadow of the Earth (Fig. 2.13). Lunar eclipses can occur only at full Moon and when the Moon is close to or at a node. The Earth's umbral shadow extends beyond the orbit of the Moon and so three kinds of lunar eclipse are

Fig. 2.12 The shadow cast by the Moon (not to scale)

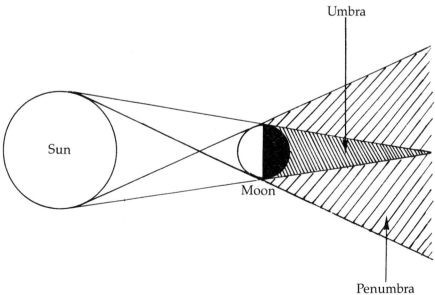

possible. A total eclipse occurs if during its passage, through the shadow the Moon becomes fully immersed in the umbra, a partial eclipse if it is only partly covered by the umbra and a penumbral eclipse if it passes through the penumbra (Fig. 2.14).

At the Moon's distance the average diameter of the Earth's umbra is about 5705 miles, which is more than double the Moon's diameter. Hence total lunar eclipses will be more frequent than solar eclipses and penumbral eclipses even more so. If the Moon passes centrally through the Earth's shadow, the average time of eclipse will be about 3 hours 46 minutes. Of this, 1 hour and 43 minutes will be total. The precise times will be affected by whether the Moon is at perigee or apogee, the Earth at perihelion or

Fig. 2.13 Eclipse of the Moon

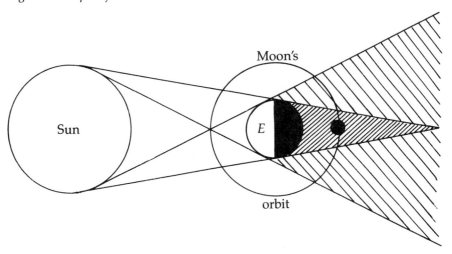

Fig. 2.14 Types of lunar eclipse (not to scale) Total, partial and penumbral eclipses

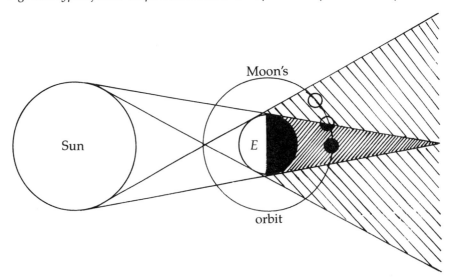

aphelion (which affects its orbital speed) and by the Moon's orbital speed. Because of the size of the Earth's shadow the Moon will pass through it and be eclipsed more frequently than the Moon will appear to pass across the Sun's disc as seen from anywhere on the sunward side of the Earth. Hence lunar eclipses are bound to be more frequent than solar eclipses. This is made even more so by another fact; an eclipse of the Moon can be seen from any part of the Earth's surface where the Moon is above the horizon whereas a solar eclipse, especially a total eclipse, is visible from only a narrow zone of the Earth's surface lying within the Moon's shadow. Therefore overall, as seen from a given observing station, lunar eclipses are much more frequent than solar eclipses.

Solar and lunar eclipses occur in a regular repeating pattern that makes them accurately predictable. For a solar or lunar eclipse to occur, the line of nodes must pass through the Earth and the Sun and the Moon must be at least near one or other of the nodes. We have previously noted that the nodes shift their position and move in a westward direction, the regression of the nodes, and take 18.61 years to make one complete revolution. Twice during this period the lines of nodes will pass through the Earth and Sun. Now, the lunar phases are repeated at intervals of about 29.53 days, the Synodic month, and the Moon passes through a given node, ascending or descending at intervals of about 27.21 days, the so-called Draconic or Nodical month (this is slightly less than the moon's true orbital revolution time as the Moon moves in a direction opposite to that of the movement of the nodes, which move toward the Moon and meet it during orbit, thus making the Nodical month shorter than the Sidereal month). It follows, therefore, that a given phase will occur again at the same node after a time interval that is a common multiple of the Synodic and Draconic months. In fact, 223 Synodic months are equal to very nearly 242 Draconic months. This is a period of about 18 years and 10¼ days and is known as the *Saros*. It has been known since 1200 BC and was used by the Chaldean priests who discovered it to forecast eclipses. After the lapse of a Saros, an eclipse seen at a given locality will be seen again. However, owing to minor pertubations of the Moon's orbital motion, the shape changes in its orbit caused by the gravitational pulls of other bodies in the Solar System and the imperfect accuracy of the Saros, a partial eclipse may be followed by a total eclipse or the other way around.

Further Reading

Books

An Introduction to the Study of the Moon. Kopal, Z. Reidel, Dordrecht, Holland (1966).

The Old Moon and the New. Firsoff, V.A. Sidgwick and Jackson, London, England (1960). Chapter 3.

The Moon. Proctor, R.A. Longmans, Green and Co., London, England (1878). This book has been out of print for very many years but should be found in most comprehensive astronomical libraries. Copies may still be obtained occasionally in antique book shops. Much of the book deals with the motions of the Moon.

3

The Moon Observer's Telescope

The telescope is, of course, the primary instrument used by the lunar observer. There are many different kinds of telescopes, each having advantages and disadvantages with respect to optical characteristics and performance, portability, ease and convenience of use and affordability. A review of the principal types of telescopes should help decide what is the best for the lunar observer with regard to the above criteria.

Refracting telescopes

Everyone knows the common spy-glass type of telescope. In its simplest form it is made from a convex spectacle lens, the object glass or objective, with a focal length of several inches, mounted at one end of a cardboard tube which is pointed towards the object being viewed and a smaller lens, the eyepiece, of much shorter focal length and at the other end of the tube, through which the observer looks. The eyepiece is mounted in a separate tube that slides in and out of the main tube so that the telescope can be focused to give a clear enlarged image of the object (Fig. 3.1). The nearly parallel light rays from the distant object are bent or refracted by the objective and converge at its focus to form an inverted image of the object. This is magnified by the eyepiece, which acts as a magnifying glass. This kind of telescope in which the optical system bends or refracts light rays to produce an enlarged image of the object is called a *refracting* telescope or simply a refractor. The magnifying power is equal to the focal length of the objective divided by the focal length of the eyepiece. Thus, a telescope with an objective of focal length 20 inches and an eyepiece with focal length of one inch will give a magnification of 20/1 (20 times), usually written 20×. (There is a widespread habit among astronomers and other telescope users to pronounce expressions like 20× as 'twenty ex', presumably because the multiplication or 'times' sign looks like a letter 'x'. The practice is incorrect and nonsensical and should be avoided. 20× is pronounced 'twenty times'.)

This simplest kind of refractor suffers from two inherent optical defects. First, the image formed by the single lens objective is marred by fringes of spurious rainbow-like colours caused by dispersion of white light as it passes through the lens, which acts like a prism. This defect is called *chromatic aberration* (Fig. 3.2).

Second, the image is never really sharp however carefully the telescope is focused because the light rays do not all come to a common focus. This is called *spherical aberration* (Fig. 3.3).

Both of these faults can be simultaneously greatly minimised by constructing the objective from two lenses, one a convex lens and the other a concave lens of a different kind of glass. Such an objective is said to be *achromatic* (Fig. 3.4). It is designed to minimise spherical aberration also and produces a sharp almost false color-free image. All refracting telescopes, except the cheapest childrens' toys, are made on the achromatic principle. Achromatic objectives are usually constructed so that their focal lengths are between 10 and 20 times the diameter of the objective, i.e., with *focal ratios* of between F/10 and F/20, as they are symbolised. Most astronomical refractors have objectives with focal ratios around F/15.

Fig. 3.1 Simple refracting telescope showing path of light rays

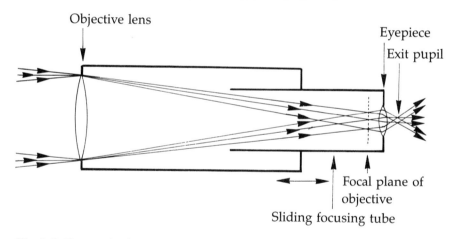

Objective lens

Eyepiece

Exit pupil

Focal plane of objective

Sliding focusing tube

Fig. 3.2 Chromatic aberration in a simple convex lens

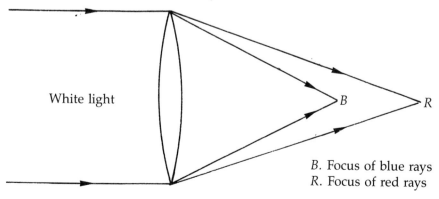

White light

B

R

B. Focus of blue rays
R. Focus of red rays

For similar reasons, the eyepieces of astronomical telescopes are made of two or more lenses and are designed to reduce the spherical and chromatic aberrations. Different types of eyepieces are available in which the aberrations are corrected to varying degrees, and these will be described later.

The object viewed with an astronomical telescope appears inverted, as may be inferred from Fig 3.1. This should not be troublesome for astronomical work as there is no 'right way up' for celestial bodies. The eyepieces of telescopes used for terrestrial viewing have additional lenses that make the object appear 'right way up'. These extra lenses would be undesirable in astronomical observation where very faint objects are being studied as they absorb a certain amount of light.

Fig. 3.3 Spherical aberration in a simple convex lens

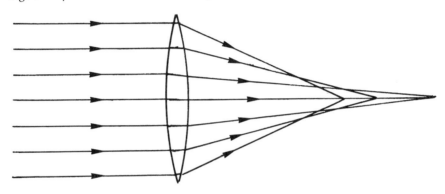

Fig. 3.4 Construction of an achromatic objective, showing how chromatic aberration is reduced by the doublet construction

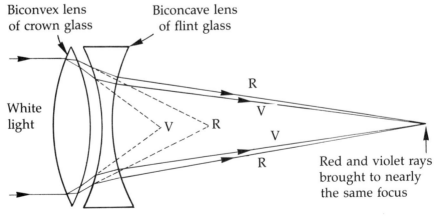

Biconvex lens
of crown glass

Biconcave lens
of flint glass

White
light

R

V

V

R

Red and violet rays
brought to nearly
the same focus

(The distance between the lenses is exaggerated)

Reflecting telescopes

A spheroidal concave mirror produces an image of a distant object (Fig. 3.5) and telescopes constructed with such a mirror instead of an objective lens are therefore called *reflecting telescopes* or reflectors. It is impossible to view the image directly with an eyepiece because one's head would obstruct the light reaching the mirror. The image must be diverted to where it can be conveniently viewed. In one type of reflecting telescope, a small plane mirror arranged diagonally within the focus of the main mirror is used to divert the converging light rays through a hole in the side of the telescope tube to the eyepiece (Fig. 3.6). This is the *Newtonian* design of reflector and is named after Sir Isaac Newton, who invented it. It is a simple and deservedly popular design for a reflector.

The advantage of the reflecting telescope is that the image formed by the concave mirror is free of chromatic aberration; however, if the mirror is spheroidal the image is marred by spherical aberration. This is negligible if the focal ratio is F/20 or greater but the telescope tube may then be inconveniently long if the mirror diameter is more than six inches. A paraboloidal concave mirror is free from spherical aberration and can be made with focal ratios as short as F/6 to F/8, so that another advantage of reflecting telescopes (if made with paraboloidal mirrors) is that they can be more compact than refractors of the same objective size or *aperture*. An eight-inch reflector (i.e., one with a mirror eight inches in diameter) is portable but even a six-inch refractor, let alone an eight-inch, is definitely not portable.

Fig. 3.5 Image formation by a spheroidal concave mirror

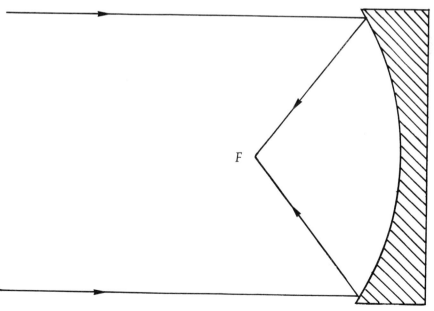

F = focus

Fig. 3.6 Newtonian reflecting telescope

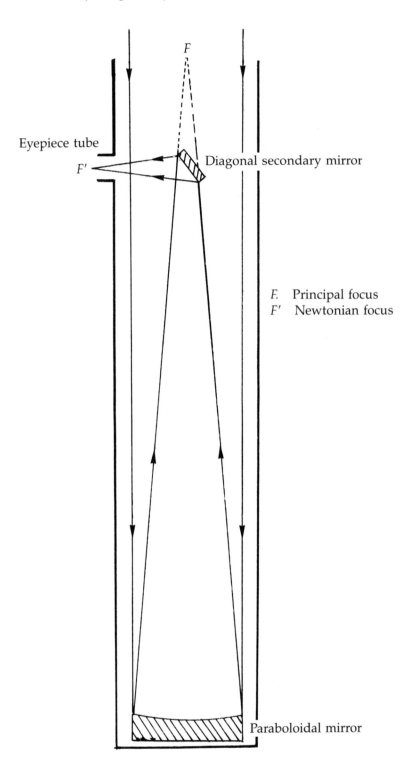

Eyepiece tube

F

F'

Diagonal secondary mirror

F.   Principal focus
F'   Newtonian focus

Paraboloidal mirror

Even more compact reflectors are those constructed on the *Gregorian* or *Cassegrainian* design. In the Gregorian, a small concave mirror placed just outside the focus of the main mirror reflects the convergent light rays back towards the main mirror and through a central hole where the eyepiece is placed (Fig. 3.7). The object being viewed is seen 'right way up'. Many years ago small Gregorians were therefore popular for terrestrial viewing but they are quite obsolete nowadays. In the Cassegrainian reflector, a convex mirror placed just inside the focus of the main mirror sends the light rays back through a hole in the main mirror to the eyepiece (Fig. 3.8). The object is seen inverted. The effective focal length is therefore much greater than that of the main mirror in both the Gregorian and Cassegrainian telescopes. They are even more compact than a Newtonian of the same F-ratio because of the 'folding' of the light path.

The Cassegrainian design and its modifications are popular for large observatory instruments.

A disadvantage with reflectors is the open tube design, which allows dust and atmospheric pollutants to fall on the mirror, which must be cleaned periodically. Air currents within the tube may also have a bad effect on image steadiness.

Catadioptric telescopes

In recent years telescopes in which both lenses and mirrors are combined have come into prominence. They are called *catadioptric* telescopes. The trouble with reflecting telescopes of the type we have been discussing is that paraboloidal mirrors suffer from an inherent aberration called *coma* which is especially noticeable at short focal ratios and so a limit is set on the compactness of reflectors. Coma affects the off-axis images of point light sources, such as stars, and manifests itself as a fan-like spreading out of the light on the side of the point image away from the centre of the field of view. The image of a star therefore looks like a comet, hence the name 'coma' for this aberration. The useful field of view in a reflecting telescope is therefore quite small. This is a serious nuisance in work where it is desired to study and photograph wide star fields but it is not too important to the lunar and planetary observer.

Spheroidal mirrors are free of coma. In the *Schmidt–Cassegrain* design of catadioptric telescope a spheroidal concave mirror is coupled with a special correcting lens that slightly alters the directions of the incoming parallel light rays before they hit the main mirror. When the light is reflected from the main mirror, having first passed through the correcting plate, it converges to form an image free of spherical aberration. A small convex mirror mounted inside the focus of the main mirror on the rear surface of the correcting plate reflects the light back through a hole in the main mirror to the eyepiece as in the original Cassegrain design (Fig. 3.9). The correcting lens or plate has a rather complex aspherical surface and is difficult to figure accurately.

In the *Maksutov* design the correcting plate has only spherical surfaces and so is much easier to produce than the aspheric corrector in the Schmidt–

Fig. 3.7 Gregorian reflecting telescope

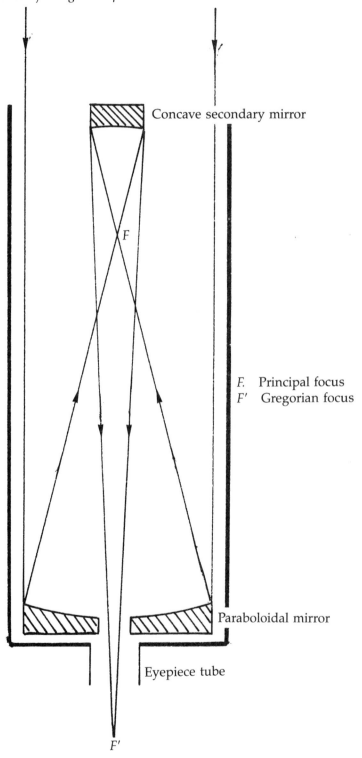

Concave secondary mirror

F. Principal focus
F' Gregorian focus

Paraboloidal mirror

Eyepiece tube

Fig. 3.8 Cassegrainian reflecting telescope

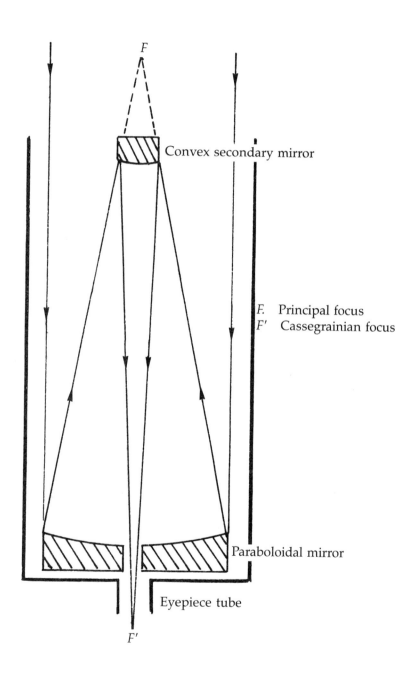

F

Convex secondary mirror

F. Principal focus
F' Cassegrainian focus

Paraboloidal mirror

Eyepiece tube

F'

Fig. 3.9 Schmidt–Cassegrain telescope

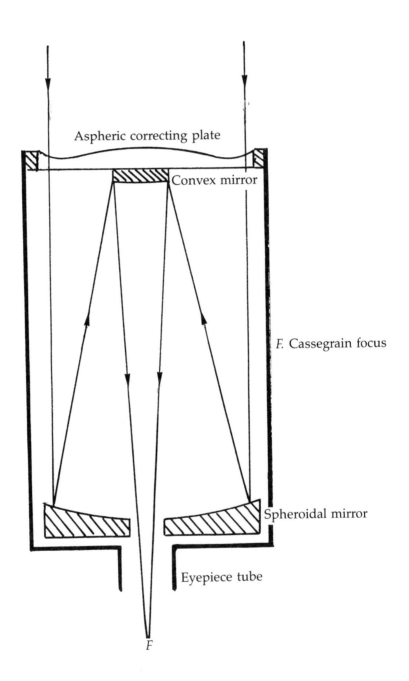

Aspheric correcting plate

Convex mirror

F. Cassegrain focus

Spheroidal mirror

Eyepiece tube

F

Cassegrain. However, it has to be quite thick and slabs of optical glass of the required thickness and free of defects are expensive. Maksutovs for normal astronomical work have the correcting plate with the convex curve towards the main mirror. The corrector can be placed within the focus of the mirror and if a small convex mirror is placed on the rear surface of the correcting plate, this gives a *Maksutov–Cassegrain* telescope (Fig. 3.10). If the corrector is placed with the concave surface towards the mirror and placed beyond the focus, a small central disc on the concave surface of the corrector may be silvered and this results in a *Maksutov–Gregorian* telescope giving erect images and much better in all ways than the original Gregorian (Fig. 3.11). Maksutov telescopes yield exquisite colourless, aberration-free images.

The main advantages of catadioptric telescopes are their wide coma-free fields and high photographic speeds, which can be as fast as F/1.5. In addition, the closed tube design protects the main mirror from atmospheric pollutants and eliminates the air currents that are often a nuisance in the open tube of ordinary reflectors, as mentioned earlier, and the design is very compact.

Resolving power and light grasp of telescopes

The ability of a telescope to reveal fine detail in the object being viewed is called resolving power. This is directly proportional to the diameter of the objective (lens or mirror). A telescope with a mirror or lens objective six inches in diameter will therefore reveal detail twice as fine as a three-inch telescope.

The surface area of the objective lens or mirror determines the light grasp of the telescope and therefore image brightness. The light grasp is proportional to the square of the objective diameter; a six-inch has four times the light grasp of a three-inch.

The Moon observer is more concerned with resolving power than with light grasp. The Moon is very bright and sometimes the telescopic view is so dazzling that optical filters must be used to reduce glare if the telescope is much over eight inches in aperture.

Eyepieces or oculars

All good telescope eyepieces consist of two or more lenses and there are many different kinds. Those most likely to be of interest to the Moon observer are as follows.

The *Huygenian* is the simplest of the two-lens eyepieces and is made from two plano-convex lenses with their convex surfaces facing the telescope objective (Fig. 3.12). The lens nearer the objective, the field lens, is placed within the focus of the objective and effectively greatly shortens the focal length of the objective, producing a new image inside the eyepiece between the two lenses. The reduced-size image is then strongly magnified by the eye lens. Between the two lenses and at the focus of the eye lens is placed

Fig. 3.10 Maksutov–Cassegrain telescope

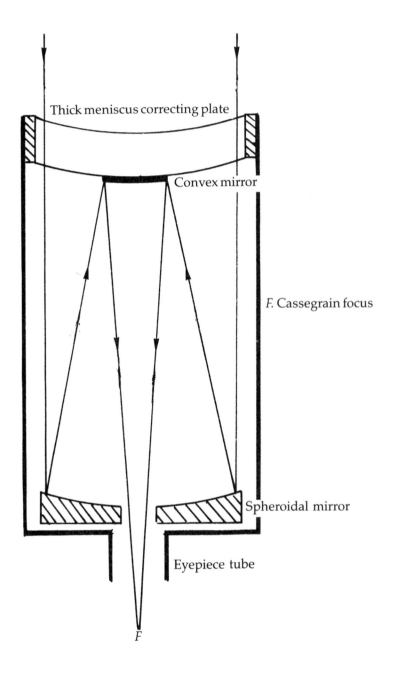

Thick meniscus correcting plate

Convex mirror

F. Cassegrain focus

Spheroidal mirror

Eyepiece tube

F

Fig. 3.11 Maksutov–Gregorian telescope

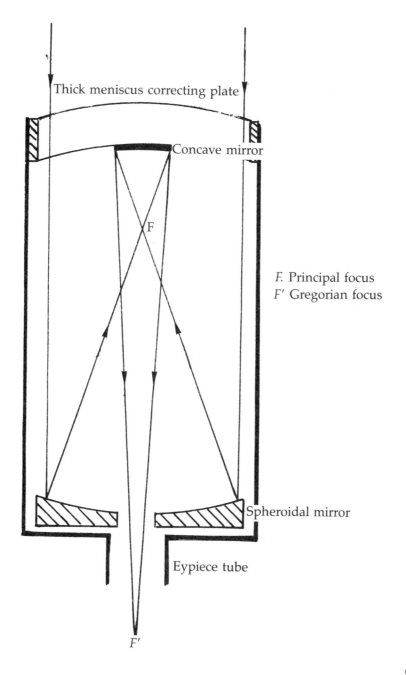

Thick meniscus correcting plate

Concave mirror

F

F. Principal focus
F' Gregorian focus

Spheroidal mirror

Eypiece tube

F'

a diaphragm with a circular hole. This is called the field stop. It gives a sharp circular boundary to the field of view.

The Huygenian eyepiece is often referred to as a 'standard astronomical eyepiece' by telescope manufacturers. It is suitable for use with refractors of focal ratio around F/15 but quite unsuitable for use with F/6–F/8 reflectors since it is not designed to handle the wide-angle light cone from such relatively short focus mirrors.

An improvement on the Huygenian eyepiece is the *Ramsden*, which consists of two plano-convex lenses with the convex surfaces turned inwardly towards each other (Fig. 3.13). Spherical aberration is much better corrected than in the Huygenian. The Ramsden can be used with reflectors with focal ratios of F/6–F/8.

The *Kellner* eyepiece resembles the Ramsden except that the eye lens is an achromatic doublet (Fig. 3.14). It is an improvement on the Ramsden but for one disadvantage – when bright objects are being viewed, internal reflections give rise to annoying 'ghost' images. The focal plane is close to the field lens and so every particle of dust on the field lens is sharply focused in the field of view. This can be very annoying in lunar observation.

Under the title of *orthoscopic* eyepieces are numerous designs. The true Zeiss orthoscopic consists of a cemented triplet field lens and a single eye lens (Fig. 3.15). It gives beautiful image quality right up to the edge of the

Fig. 3.12 Huygenian eyepiece

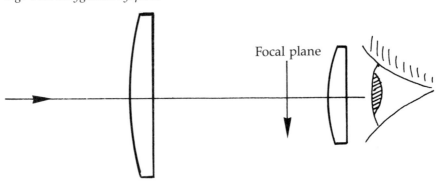

Focal plane

Fig. 3.13 Ramsden eyepiece

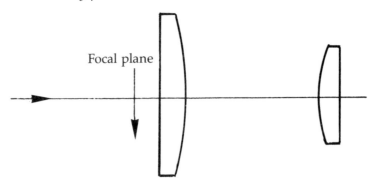

Focal plane

field of view. Orthoscopics are the best type of eyepiece to use with reflectors – but they are expensive.

Nagler and *Plössl* eyepieces are of complex design and very expensive. Much is claimed for them by manufacturers regarding image quality and field diameter. The moon observer may wish to try them if money is no object but orthoscopic eyepieces will be found to be excellent and much less expensive.

Eyepieces for astronomical telescopes are always mounted in cylindrical metal tubes usually 1½ inches in diameter. Some telescopes of Japanese make are made to take eyepieces 0.965 inches in diameter. The size of the eyepiece makes no difference to optical performance and either type may be used with any one telescope using appropriate adapters which are commercially available. Eyepieces are designed to slide into the eyepiece holder of the telescope.

The Barlow lens

This is a valuable accessory for reflector users. A six- or eight-inch reflector of focal ratio around F/7 has about the same focal length as a F/15 three- or four-inch refractor. Shorter focal length eyepieces must therefore be used to obtain high enough magnification to exploit the resolving power of a six- or eight-inch mirror. Such eyepieces may be awkward to use as they have small eye relief, i.e., the eye has to be placed inconveniently close to them. The *Barlow lens* is a negative (diverging) lens that is placed just within

Fig. 3.14 Kellner eyepiece

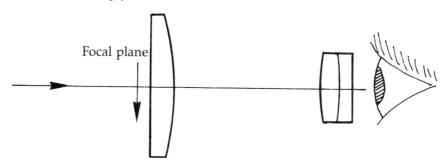

Fig. 3.15 Zeiss orthoscopic eyepiece

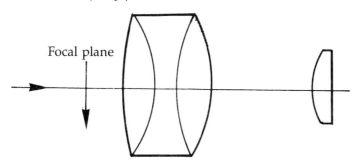

the focus of the telescope mirror and has the effect of reducing the angle of the light cone (Fig. 3.16). This is equivalent to greatly increasing the effective focal length of the telescope while adding only an inch or two to the actual overall focal length of the mirror. The image produced is therefore much larger and eyepieces of longer focal length, which are more comfortable to use, will yield higher magnifications. Also, since the light cone has a smaller angle, any eyepiece will perform better with than without the Barlow lens. Amplification provided by Barlow lenses is commonly 2–3×.

Now that a brief overview of the theoretical side of telescopes has been given, let us turn now to the practical matter of choosing the best type of telescope for the Moon observer.

Buy or build?

High quality astronomical telescopes, mostly reflectors, some of them quite large, have been built and are being built all the time by hundreds of dedicated amateur astronomers. All that telescope building requires apart from a flair for 'making things' are patience, perseverence and the ability to stick to a project until it is finished even though plagued with difficulties and setbacks. There seem to be two types of telescope enthusiasts, 'makers' and 'users'. The 'makers' delight in constructing telescopes and the only time they seem to look through the eyepiece is to test the perfection of the telescope mirror with some kind of natural or artificial test object. When it is judged that the mirror has passed all tests, they pass on to the next mirror-making project.

Fig. 3.16 Action of the Barlow lens

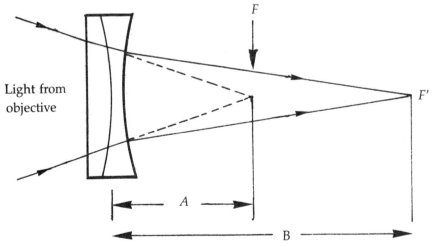

F = focus of objective
F' = focus with Barlow lens

Here, the distance B equals 2A, therefore the amplification factor of the Barlow lens is 2×

Now, I admire such people and envy them their expertise but am not myself one of them; admittedly I have never even tried my hand at constructing a serviceable telescope mirror. If any of my readers want to attempt to make their own telescopes, there are supplies of telescope making kits and mirror 'blanks' available at modest cost but it is beyond the scope of this book to tell you how to build your own telescope and furthermore, as I have remarked, I have no experience at it. There are plenty of good books on telescope making and I suggest that the reader consults the list at the end of this chapter if he wishes to make his own.

Refractor or reflector?

If the reader is like myself – a 'user' – he will want to buy his telescope ready made. Telescopes, even small ones, can be quite expensive and so the questions of what sort of telescope to purchase – refractor, reflector or catadioptric – and how much money will be required are important ones .

There are at least two characteristics of the telescope that are of paramount importance if you are an amateur with a moderate income like myself and who will probably never aspire to ownership of a permanent and beautifully appointed observatory equipped with revolving dome and a large telescope. These characteristics are: (1) price, (2) portability. Since you will want to purchase the largest possible aperture for your money (if you intend doing serious work) and at the same time have an instrument that is reasonably portable, I make my recommendation without any hesitation; buy a Newtonian reflecting telescope from six to eight inches in aperture.

A Newtonian reflector costs far less than a refractor of similar aperture and it is portable. A three- or four-inch reflector is not much use to the serious lunar observer and a ten- or twelve-inch reflector is not really portable. Standard six- and eight-inch refractors are not portable and are definitely observatory instruments. Such telescopes are not often found in private amateur hands and are enormously expensive.

Much has been written on the relative merits and demerits of refractors and reflectors, the two basic types having certain inherent advantages and disadvantages, both practical and theoretical. I do not intend going into a full discussion of these here except to mention that the strong point of the reflector is that the image produced by a paraboloidal mirror is completely color-free or achromatic whereas the image produced by the achromatic objective of a refractor still has some residual color. On the other hand, the refractor objective is not obstructed by the secondary mirror and its supports as in the Newtonian telescope; these can have detrimental effects on image contrast and result in some loss of light, but the latter is of no consequence to the lunar observer. There are other lesser considerations regarding compactness and portability, ease of housing, ease of maintenance and keeping in adjustment, relative comfort in using the two types of telescope and open tube (reflectors) versus closed tube (refractors). The choice of telescope involves compromise and, in my experience, the best value for money – where that is limited – and portability is undoubtedly the reflector of the aperture range just mentioned. For many years I was privileged to use the eight-inch refractor of the Kellog Observatory, Buffalo Museum of Science, New York, USA, to study the Moon before I purchased my own eight-inch reflector. I never noticed any appreciable difference in performance nor did I sense any disappointment when I started using my reflector after so many years of using the eight-inch refractor. In fact, I believe that modern reflector mirrors have reached such high standards that to all intents and purposes their performance is equal to that of a refractor of equal aperture, certainly in lunar observation. Refractor enthusiasts will disagree with me no doubt and would recommend a four- or five-inch refractor over a six-inch reflector. I started lunar observing with a three-inch refractor and it was many years before I could afford to buy my eight-inch reflector. The larger aperture, even over a four-inch, reveals considerably more lunar detail than the smaller refractor. However, I would never part with my three-inch. It gives superb views of the Moon for casual 'moon gazing' and for showing the Moon to interested friends it is fine and far easier to set up than the eight-inch. However, I doubt whether much lunar work of real scientific value can be done these days with instruments of four inches aperture or less.

I would not recommend one commercial telescope maker's product over that of another – it would not be in any manufacturer's interest to produce optics of inferior quality.

In the catalogs and advertisements of reflecting telescope manufacturers, you will often see it stated that the paraboloidal surface of the primary mirror has been shaped – or *figured*, to use the correct term – to 1/8th wave accuracy. This means that the figure of the concave paraboloidal surface does not differ from theoretical perfection by an amount more than 1/8th of the wavelength of light, i.e., about three millionths of an inch. Mirrors are tested for accuracy of figure by an optical test known as the Foucault test, which though relatively simple is capable of revealing incredibly small deviations from perfection in astronomical mirrors. Without going into details, the principle of the test is that it enormously magnifies imperfections in the mirror shape by optical means and makes them easily visible. Amateur mirror makers use this test frequently during the final figuring of mirrors to ascertain when the mirror surface is correctly shaped. Sometimes, even 1/10th and 1/20th wave accuracy will be advertised; however, the difference in performance between a 1/8th wave mirror and a 1/20th wave mirror will only usually be perceptible to an experienced observer. The improved performance is noticed not so much in resolving power as in improved contrast in lunar and planetary detail. Good seeing and high magnification are necessary for the improved performance to be apparent. The lunar observer will be well satisfied with a 1/8th or 1/10th wave mirror.

In a reflector of F-ratio greater than F/10, the shape of a paraboloidal mirror does not differ from that of a spheroidal mirror by more than the 1/8th wavelength limit so that long focus reflectors can be made with spheroidal surfaces.

Because inaccuracies in the figure of reflecting telescope mirrors are magnified by the effect of the reflected light, reflecting telescope mirrors generally have to be figured much more accurately than objective lenses.

What about catadioptrics?

When I was thinking of buying a telescope larger than my three-inch and had decided that I wanted an eight-inch reflector, I was for a time attracted to the Celestron series of telescopes, one of which is an eight-inch. The Celestron is a Schmidt-Cassegrain design and the Celestron–8 (the eight-inch model) is wonderfully compact and portable; it can be easily carried with one hand. The price of the Celestron–8 was double that of the corresponding Newtonian eight-inch as supplied by most manufacturers. I reasoned that the Celestron–8 would not reveal any more fine lunar detail than a Newtonian eight-inch, so that the portability of the former was rather an expensive luxury. Also, the Schmidt–Cassegrain is really designed to yield wide-angle flat fields for stellar photography. The rather large central obstruction would result in some light loss but worse still I was rather worried about the effect on contrast of lunar and planetary images. Maybe I exaggerated this to myself. I only ever tried out a Celestron–8 on two occasions on the Moon but conditions were such that I could not make a fair comparison between the Celestron and a Newtonian, even if the latter had been available at the time. The more I saw that large central obstruction the less I liked it and rightly or wrongly I felt that the Celestron's strength might lie in areas other than lunar or planetary observation. I finally purchased an eight-inch Newtonian and have never regretted it. I also saved a lot of money. Generally speaking, Cassegrains of the usual (classical) sort are best left alone by the amateur. One author has stated that they have many of the disadvantages of both refractors and reflectors. Maksutovs are very expensive even in the smaller sizes but they give exquisite images.

Your first telescope

Whether it is best to start lunar observing with a small refractor (2- or 3-inches) and then 'graduate' to a six- or eight-inch reflector, as I did, is a matter for debate. In the early stages , the views of the Moon provided by a three-inch refractor , if one has only ever been used to binoculars, is quite breathtaking. There is sufficient lunar detail revealed by a three-inch to keep you observing, studying and drawing for a lifetime. If you decide, at length, that the Moon is not for you, then you will not have spent too much money and you will have a nice convenient telescope for general star gazing instead. If you are a real lunar enthusiast, you will soon feel the need for something larger and will want a six- or eight-inch reflector. Your three-inch will still come in handy, as I have previously mentioned.

One may equally argue for going to the six- or eight-inch reflector straight away; why waste time with a three-inch? Quite so. However, I personally believe that it is better to train the eye and to obtain initial observational practice with a three-inch; the greater magnification and the better and more detailed views of lunar features afforded by the larger telescope are then more fully appreciated; learning to walk before running, so to speak.

If at first you decide to purchase a small refractor of two or three inches aperture for financial or other reasons, do not be mesmerised by the extravagant claims made by some manufacturers of enormous magnifying powers. There is such a thing as 'empty' or 'useless' magnification, which will be explained later in this chapter. Sometimes magnification is stated in terms of 'areas', for example a given telescope will be stated to magnify to '400 areas'. This means that the area of the object being looked at appears to be 400 times greater than with the naked eye. This sounds very impressive until it is realized that the 400 areas is simply the square of the actual linear magnification, which in this case would be 20 times, which is really quite modest. Another thing to guard against is being attracted by long lists of accessories and gadgets. A two-inch refractor elaborately mounted and accompanied by an array of accessories and a handsome carrying case is still only a two-inch refractor and will show only what a two-inch refractor will reveal. It is far better to invest the money that you would spend on the gadgetry and use it to buy a three-inch simply mounted and without any gadgets, at least at first. These can be added later as funds permit.

Eyepieces

As soon as you have acquired any telescope the question of what eyepieces to buy will arise. Let us assume that you have purchased a reflector; do not buy any eyepiece advertised as a 'standard astronomical eyepiece'. Impressive though this may sound, these are really ordinary Huygenians and are suitable for use only with long focus refractors and are quite unsuitable for reflectors. To use a Huygenian eyepiece with a reflector is equivalent to playing a record album with a very blunt stylus – it is simply impossible to get the best results with such a combination. Ramsdens and Kellners give good results and so do symmetricals; the latter type of eyepiece consists of two identical achromatic doublets positioned almost in contact with similar surfaces facing each other, hence the name. If you do not mind spending more money, buy the orthoscopic type. They are unbeatable; I rarely use any other type of eyepiece. If you can afford only one eyepiece at the outset, buy an orthoscopic of 12 millimeters focal length (1/2 inch). This will give a medium power of about 100 – 130× with the usual F/8 six- or eight-inch reflector. When more funds become available invest in a 2× or 3× Barlow lens – but make sure that this is a good achromatic design such as the Goodwin or Dakin type. Avoid the cheaper non-achromatic Barlows that are sometimes advertised, usually of Japanese make, for use with small refractors. Coupled with your 12 mm orthoscopic eyepiece this will give you the highest power that you are likely to need and you will be able to exploit the full resolving power of your telescope. Later, a low power eyepiece, say a 25 millimeter (1 inch), can be added. Used alone this will be your lowest power and with your Barlow this will give you another higher power. Assuming that you have a 25 mm and a 12 mm eyepiece and a 3× Barlow, the powers that you will obtain on a F/7 eight-inch reflector will therefore be:

	alone	with 3× Barlow
25 mm	56×	168×
12mm	112×	336×

Be careful not to duplicate the powers of some of your eyepieces when purchasing a Barlow lens. For example, if you possess 24 mm and 12 mm eyepieces, and you avail yourself of a 2× Barlow, this when used with the 24 mm eyepiece will merely duplicate the power obtained with the 12 mm alone. It is best to find a Barlow with a non-integral amplification factor such as the Vernonscope 2.4× or the Edmund RKE 2.5×. Duplication of powers cannot then occur. To achieve high powers it is better to purchase a Barlow lens and to use it in conjunction with a relatively lower power eyepiece. By so doing you will avoid one of the main faults with very high power eyepieces such as those of 6 mm and 4 mm focal length, which is their very short eye relief. This means that you have to put your eye uncomfortably close to the eyepiece in order to see the full field of view. This is a real nuisance if you are a spectacle wearer. On the other hand, a 12mm eyepiece with a 2× or 3× Barlow lens will give the magnification equivalent of a 6 mm or 4mm eyepiece but will be much more comfortable to use as the eye relief will be the same as that of the 12 mm, which will be better than that of the 6 mm or 4 mm eyepieces. Furthermore, the Barlow lens makes the convergent light beam reflected from the main mirror much more narrow angled and all eyepieces will perform better because of this.

The list of eyepieces and Barlow lenses given above is only one of various selections that might be made. Whatever you choose, try to have at least three power ranges at your disposal:

1   A low power of 50× to 70× for low power surveys of the Moon or for viewing lunar eclipses.
2   A medium power of around 200× to 250× for routine detailed study of lunar topography.
3   A high power of about 300× to 400× for checking doubtful detail or for when the seeing conditions are really excellent and will permit the use of such high powers.

Telescope mountings

No serious observational work can be done with a telescope unless it is properly mounted on some kind of rigid support. Try looking at the Moon with an ordinary small terrestrial refractor or 'spy-glass' held in your hand. Then mount the telescope on a rigid stand of some kind and look at the Moon again. You will be surprised at the extra detail that you can see. Vibration and movement, however slight, are always detrimental to telescopic vision. This simple experiment should convince you of the utter foolishness of attempting to hold a six- or eight-inch reflector in your hand while trying to observe the Moon. Apart from the sheer weight and size of such telescopes a good rigid mount is even more important for a six- or eight-inch reflector than it is for a two- or three-inch refractor because of the much higher magnifications that can be used. A good telescope mount should possess the following features:

1   It must be quite rigid and vibration-free.

2   The telescope tube itself must be easily moved on the mount so that the Moon may be followed as it moves from east to west in the sky, due to the Earth's rotation.

There are two principal types of mounting:

1. The altazimuth. In this type the telescope is supported in a fork-shaped structure and rotates in a vertical plane about a horizontal axis. The fork itself rotates on a vertical axis (Fig. 3.17). Two motions can therefore be imparted to the telescope, one in altitude (up and down) and one in azimuth (side to side). This is what gives the altazimuth mounting its name. Small refractors are frequently mounted in this way. The mount itself is attached to a rigid upright pillar or a heavy tripod. My own altazimuth-mounted three-inch refractor is on a large wooden garden tripod and the whole is wonderfully rigid. During prolonged observing it is a nuisance to be continually readjusting the telescope in altitude and the azimuth to keep the Moon in the field of view as it moves across the sky, especially when high powers are being used, but it can be 'lived with'. Of course, the higher the power used, the more rapidly does the Moon seem to move and the more frequently do the adjustments have to be made. It often comes as a complete

Fig. 3.17 Altazimuth mounting for a small refracting telescope

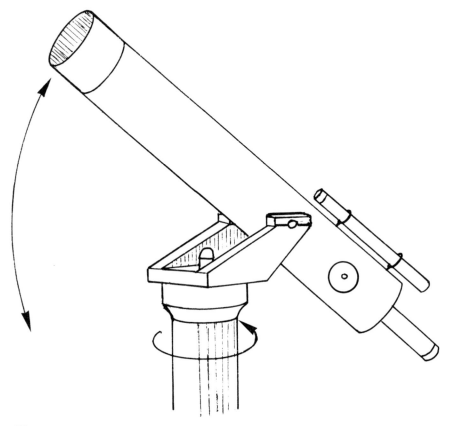

surprise to the beginner to see how quickly the Moon moves in the field of view of a telescope even with moderate magnifications. However, in spite of this drawback, the altazimuth is a good and useful mount for a small or moderate telescope. Astronomical altazimuth refractors are often supplied with hand-operated slow motion devices to both azimuth and altitude axes and the movement of the telescope needed to follow the Moon is more convenient and precise.

2. The equatorial. There are many modifications of the equatorial design but they all have one feature in common; the axis about which the telescope moves in an east–west direction is not vertical but is inclined to the ground at an angle equal to the observer's latitude. If the telescope is exactly aligned in a north-south direction, the axis will now be parallel to the Earth's axis. If the telescope is tilted about the other axis (the *Declination axis* as it is called in the equatorial mount) and pointed to the Moon, all that is necessary to 'follow' the Moon is simply to rotate the other axis (the *polar axis* as it is called since it points to the celestial poles) and the telescope will follow the Moon. Two movements are not therefore necessary as they are with an altazimuth. Modern equatorial mounts are usually fitted with an electrically driven motor that turns the polar axis in an east to west direction so as to counteract the Earth's west to east motion and at a speed that will enable the telescope to follow the Moon and keep it in the field of view for protracted periods, provided everything is accurately adjusted. A motor drive would not be feasible with an altazimuth since two movements at right angles to one another have to be made. It is true that a very large altazimuth-mounted telescope in the Soviet Union is electrically driven but the complex movements in altitude and azimuth necessary for it to be kept pointing at a heavenly body are done by a computer. I have not heard of such a device being commercially available for amateur-owned altazimuth-mounted telescopes; even if it was, the price would probably put it beyond the reach of most amateurs anyway.

The equatorial can be considered as an altazimuth in which the vertical axis has been tilted and aligned until it is parallel to the Earth's axis and points to the celestial poles.

Virtually all commercial reflectors for amateur use are made with equatorial mounts and the serious lunar observer should obtain one with an electrically driven polar axis. His hands and mind will then be free to concentrate on lunar observing and recording without being distracted by continually having to readjust the telescope to keep the Moon in the field of view.

The equatorial design ordinarily supplied is the German type (Fig. 3.18), in which the telescope tube is attached to one end of the Declination axis, its weight being balanced by counterweights attached to the other end of the axis.

The usual type of support for the equatorial reflector is the vertical pillar with three horizontal legs at an angle of 120 degrees to each other. This arrangement is quite rigid. The best and most rigid type of support is a metal pillar sunk several feet into the soil and surrounded with concrete.

However, this is not portable although there is no reason why the equatorial head itself should not be detachable so that it can be taken indoors when observing is finished for the evening.

Apart from ease and convenience of movement of the telescope tube, the mount should be quite rigid and vibration-free. However, there are not many commercial mounts in which these ideals are fully realized. The vertical pillar on three horizontal legs is quite rigid but the equatorial mount itself is often not rigid enough for vibration-free viewing. In fact, the mount would have to be so heavy and massive to achieve this, that it would add

Fig. 3.18 German equatorial mounting for a reflecting telescope

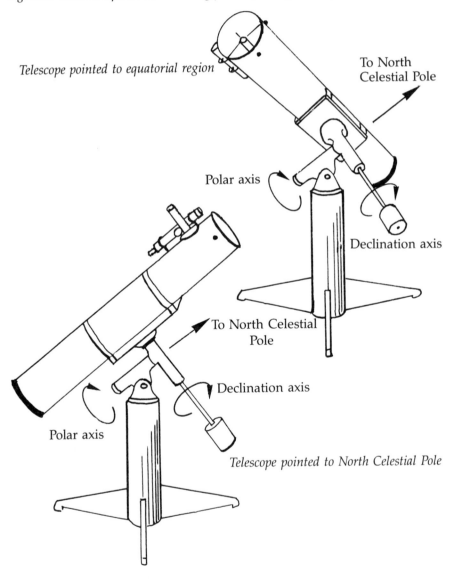

Telescope pointed to equatorial region

To North Celestial Pole

Polar axis

Declination axis

To North Celestial Pole

Declination axis

Polar axis

Telescope pointed to North Celestial Pole

considerably to the cost and would severely limit the portability of the instrument.

Setting circles

Attached to the declination and polar axis you will find metal or plastic discs a few inches in diameter that are graduated around their edges. The one attached to the Declination axis (the *Declination circle*) is graduated in degrees and that attached to the polar axis (the *hour circle* or *Right Ascension circle*) is graduated in hours and minutes. Each axis possesses a pointer so that the graduations can be read off on these *setting circles* as they are called. They are given this name because they can be used to locate or set the telescope on very small or faint celestial objects at night or planets like Venus and Jupiter during the day. If it is desired to point the telescope accurately at a celestial object that cannot be easily seen with the naked eye, the Right Ascension and Declination of the object at the particular time are calculated from the relevant data for that object. This information is listed in publications such as the *Handbook of the British Astronomical Association* and the *Observer's Handbook* of the Royal Astronomical Society of Canada. Assuming that the telescope is properly set up and that the setting circles are accurately adjusted, the object may be found by rotating the polar axis until the pointer indicates the Right Ascension of the object at that particular moment and the Declination axis is rotated until the object's declination is indicated on the circle. The object should be seen in the field of view on looking through the eyepiece. The latter should prefer-ably be of low power, so that the searched-for object will still be found in the field of view if there are slight errors (as there almost certainly will be) in the setting up and alignment of the telescope and in the adjustment of the setting circles.

For reasons that should be obvious, the lunar observer will have no use for setting circles and so they will not be discussed any further.

The observing site.

The owner of an expensive telescope will experience frustration and disap-pointment if it is set up on a site unsuitable for lunar observing. The most unsatisfactory site of all is the room in the house where you are sitting, just throwing open a window and sticking the telescope through it will give the worst possible view of any celestial object. The indoor temperature is almost certain to be different from that outside so that air currents and turbulence will be caused by the mixing of masses of air at different temp-erature. This makes the refractive properties of the atmosphere in the neighborhood of the window so unstable that the image of anything seen with the telescope will be shimmering and quivering so much that all fine detail in the image will be impossible to see. The only satisfactory observing site for the amateur lunar observer is outside the house, usually the back garden.

First and foremost, it is imperative that you should have a clear and unobstructed view of the southern part of the sky and as much to east and west of the meridian as possible. Second, there should be a firm level surface on which to stand the telescope. Third, the telescope should be as far as possible from the house and away from chimneys so as to minimise the turbulence in the atmosphere caused by heat. Fourth, it is an advantage to be sheltered as far as possible from the prevailing winds. Apart from comfort, the more important reason for shelter is that wind seems to cause most telescope mountings to wobble. This can be annoying if the conditions for observing happen to be good, such as clear sky and freedom from atmospheric turbulence. In summary, a fairly good observing site, at least in England, would be a back garden on the east side of the house sheltered from westerly winds and which provides an unobstructed view of the south part of the sky with as much open sky on each side of the meridian as possible.

The above discussion relates to your immediate local conditions and what they should be if you are to have a good observing site. Apart from these there will be the question of whether the general area in which you live is good for observing. If you live in the country, the atmosphere is unlikely to be polluted and will therefore be clear and transparent. If you live in or near an industrial area, the air is always more or less contaminated with smoke and grime and the atmosphere will be murky but this need not be the tragedy that it is sometimes made out to be, at least for lunar observing. It is really worrisome only to devotees of 'deep sky' observing, i.e., observation of star clusters and nebulae mostly, which are rather faint objects. Street lighting and other forms of 'light pollution' are also a great hindrance to deep sky work but to the lunar observer are little more than slightly irritating visual distractions. Another problem, which like these is beyond the control of the individual, is vibration caused by heavy traffic. Unless you live close to a main highway this need not be too serious either. However, if you decide that you must have perfect transparency of the atmosphere, no light pollution and complete freedom from all vibration in an area where none of these is realised, then generally the only solution is to move home to a more satisfactory location.

Housing and care of the telescope

It is not necessary to house your telescope in an expensive observatory complete with revolving dome in order to pursue serious observation of the Moon. On the other hand, to leave the telescope permanently set up out of doors without any protection from the weather would, of course, be madness, and no normal person would be so stupid.

To be able to walk into an observatory and have the telescope immediately available for observation without any setting up is a great convenience. If the reader is 'good with his hands' and wishes to build an observatory around his telescope this is fine; once again, to describe the construction of an observatory is beyond the scope of this book and the reader will find a list of books at the end of this chapter that contain information about the building of small observatories, or various types of shelters for the telescope. A domed observatory is the ideal but it is by no means essential. Two practical alternatives to the domed observatory are as follows:

1 *A simple shelter made of wood and waterproofed* having hinged top and sides that can be opened up to expose the telescope. Alternatively, the shelter can be on wheels so that it can be rolled away on improvised 'railway lines' when the telescope is needed. Angle iron serves very well for the 'railway lines'. A third possibility is for the shelter to be in two halves which are rolled away or folded back from the telescope. Although the telescope is protected from the weather by devices like these there is no protection from the wind during observation; with every puff of wind the telescope tube will vibrate exasperatingly and the lunar image will perform a fantastic dance in the field of view.

2 *A shed with a flat 'run off' roof* that can be rolled away on rails above ground extending from the end of the shed so as to expose the sky. Alternatively, the roof may be hinged so as to fold back in one or two parts. The shed with removable roof is better than a simple removable cover for the telescope because the shed gives protection from the wind to both telescope and observer as well as protecting the telescope from the weather. Moreover, there is room to house a chair and perhaps some accessories like flashlights, books and charts.

It goes without saying that one cannot be permitted the luxury of a heated observatory on cold winter nights. The air currents set up by the warm air rising would wreck the seeing and cause the lunar image to appear as though it were being viewed by reflection from the surface of rippling water.

One need not leave the entire telescope out of doors. The pillar and the equatorial head may be left outside under the shelter or in the shed (as indeed they will have to be if the pillar is permanently embedded in the ground and the equatorial head permanently attached to it), and the telescope tube taken out of its cradle and stored elsewhere. If this is the case, the telescope tube should preferably be kept in an outhouse or garage where the temperature is about the same as out of doors. If the telescope tube is brought from indoors for use on a winter evening, it should be left outside for at least half an hour for its temperature to drop to the outdoor temperature. The mirror is affected by temperature changes and its figure may be temporarily distorted during cooling, which results in poor image quality; also, air currents ('tube currents') are set up in the open tube of the reflector while it is cooling, which is detrimental to the image. In other words, it is no good trying to carry out critical lunar observation on a cold evening before allowing the telescope to 'acclimatise' to the outside temperature if brought from a warm indoors.

I myself have never had the luxury of a permanently set up pillar and equatorial head in the garden; I store my complete telescope in a garden shed. It is taken apart into three units - the tube itself, the pillar and foot and the counterweights. When about to observe I take out the pillar–foot first and set it up in the garden. (Here, I would caution that if you anticipate doing the same, be very careful not to hurt your lower back when lifting a heavy weight up from the ground and setting it down again. The pillar–foot–equatorial head assembly of an eight-inch reflector can be quite heavy. Always bend your knees when lifting – not your back – and let your legs do the work; they are much stronger and more powerful than your lower back. Keep your back perfectly straight when lifting.) Then I attach the counterweights to the lower end of the polar axis. (If you attach the telescope tube first, the polar axis will be top heavy and just as you turn your back to pick up the counterweights, the telescope tube will swing around and hit the ground.) The Declination axis is then locked so that it cannot rotate and the telescope tube laid in its cradle and the retaining rings screwed down. Then the electric drive is switched on, the declination axis unlocked and the telescope pointed to the Moon and the declination axis locked again. An eyepiece is then slipped into the eyepiece holder and focused. Thus, in about five minutes or less from commencing to set up, I can start observing. The setting up and taking down routine is no great nuisance nor a time waster. One soon gets used to it. Even this can be avoided by the simple expedient of attaching wheels to the ends of the three legs of the stand or mounting the whole telescope on some kind of cart so that the entire thing can simply be wheeled in and out of a garden shed or garage.

I have often thought of constructing some sort of shelter for the telescope or a simple observatory so that I can leave it set up outside but have never 'got around' to doing so; I have never felt lack of it to be a serious handicap, however.

Protection from dust and atmospheric pollution

One of the disadvantages of the open tube of the reflector is that it allows dust to collect on the main mirror and the diagonal, so that when the telescope is not being used it should be protected from dust. I always stand my telescope vertically in a corner of the shed and cover the open end with a close-fitting circular cardboard lid. I was very fortunate in finding one of just the right diameter. A good alternative is to cover the open end of the tube with a polythene bag and tie it tightly around the end of the telescope tube. Even a sheet of cardboard or thin wood laid over the open end of the tube would suffice. One could also stand the tube on its open end so that the mirror end is uppermost. This would cover the open end of the tube and in addition, since the mirror is now facing downwards, it would not be so easily soiled by dust falling on to it. However, the tube would now be top heavy and might topple over if given something more than an accidental nudge. This hazard can be avoided if you can find a tall narrow cupboard in which to stand the tube. Do not forget to cover the eyepiece tube and the objective of the viewfinder. Commercial reflectors are usually supplied with plastic caps for this purpose.

Atmospheric pollutants, especially sulfur dioxide and hydrogen sulfide in industrial areas, will cause a silvered mirror to tarnish rapidly, so that it will have to be dismantled periodically for resilvering. A good aluminum film should last for several years and still retain good reflectivity especially if coated with one of several types of transparent protective films that are available.

Cleaning the mirror

The sight of even a few specks of dust on the beautiful silvery surface of the main mirror will often upset the recent buyer of an expensive reflecting telescope. This is strange since there is already a relatively large obstruction in the light path (the diagonal mirror) but this does not seem to worry telescope owners. Dust on the mirror is vividly seen if you point the telescope at the Moon and then look down the open end of the tube at the mirror. However, even quite a lot of dust on the mirror has little detrimental effect on the image. There is only very slight loss of light and this will not usually be a problem to the lunar observer. Far more harmful are greasy streaks, films or fingerprints. I had my own telescope for seven years before I cleaned the mirror and even then its necessity was doubtful. Some observers seem to be unperturbed even when the mirror looks like the bottom of a garbage can. If you decide to remove the telescope mirror to clean it, do not wipe it with a cloth; this could very easily cause scratching of the surface. Stand the mirror in warm water to which a little good mild detergent has been added. Swab the surface gently with cotton wool soaked in the water. Rinse the mirror in tap water and finally in distilled or deionized water and then let it stand on edge to dry. There should be no streaks or marks left on the mirror when the water has evaporated. If there are, wet the mirror again and try a final rinse with alcohol if you can get it. Do not use ordinary methylated spirit but industrial methylated spirit that does not give a cloudiness with water. To obtain this you will need to write to your local Customs and Excise Office asking them to supply you with a special permit which you will need before your order for industrial spirit can be accepted by a supplier.

Generally speaking, do not be anxious to remove the mirror frequently for cleaning. There is always some risk of inadvertently scratching or otherwise damaging the surface. If the mirror looks a bit dusty but seems to be performing well, then it is best to leave well enough alone.

Collimating the telescope

If an optical instrument such as the telescope is to function properly, all of its optical components must be in correct alignment and properly 'squared on', otherwise the image quality will be more or less poor. Small refractors up to about five or six inches of aperture do not usually have means for adjusting the alignment of the objective and eyepiece as these are permanently and correctly set by the manufacturer. Once adjusted they usually 'stay put'.

In a reflecting telescope, the optical components are adjustable and before the telescope can be used the alignment of the optical components should be checked. If, as is often the case, the main mirror is misaligned, it has to be brought into correct alignment with respect to the diagonal mirror and the eyepiece. The process of bringing the mirror into correct alignment with the other optical components is called *collimation*.

The diagonal mirror should be exactly on the optical axis of the main mirror and its surface should be at an angle of 45 degrees to it. The eyepiece tube should be exactly at right angles to the optical axis of the main mirror and the axis of the eyepiece holder should pass through the exact centre of the diagonal mirror (Fig. 3.19).

In practice, the diagonal mirror will be fixed accurately by the manufacturer in its proper position with respect to the eyepiece tube and the optical axis of the main mirror. It is the main mirror only that ordinarily needs adjustment; it is easily put out of alignment by undue vibration or knocking and so the telescope tube should always be handled very gently.

To test the collimation of the telescope, set the telescope up with counterweights attached and tilt the tube so that it points to the open sky or any other bright surface. Remove the eyepiece and look through the eyepiece tube exactly along its optical axis. This can be made easier by constructing a peep sight from the snap-on cap of a discarded plastic 35 mm roll film canister. A small hole 1 – 2 mm in diameter is pierced exactly through the center of the cap, which is then slipped over the end of the eyepiece tube. On placing the eye close to the hole, you will now be looking accurately along the optical axis of the eyepiece tube. What you should see is the apparently circular form of the diagonal mirror (it is elliptical but looks circular as it is tilted at an angle to the eyepiece tube) and in it reflected the circular image of the main mirror, nearly but not quite filling the circular outline of the diagonal. The image of the diagonal should be exactly in the centre of the main mirror image and you should see an image of your eye within the image of the diagonal with the pupil exactly central. Radiating out from the diagonal image to the circumference of the main mirror image will be the spoke-like reflections of the struts that support the diagonal mirror (Fig. 3.20). If the main mirror is out of adjustment , and it probably will be, you will see something like the appearance in Fig. 3.21.

To get the mirror into correct alignment, locate one of the three adjusting screws in the main mirror mount and give it a quarter turn and then look into the eyepiece tube again. If matters are worse, turn the screw back again; if better, leave this screw alone and go on to the next one. Do the same and go on to the third screw. Continue adjusting the screws until the appearance of Fig. 3.20 is seen. The telescope will now be properly collimated. It is convenient if you can get someone to adjust the screws for you as you look into the eyepiece tube so that you can give instructions while you watch the progress of collimation; however, it is not too tiresome to do it on your own and with some experience the adjustment need not take more than a minute or two.

Collimation by this method may be further refined, if desired, by inserting a high powered eyepiece and studying the appearance of a star image and adjusting the main mirror cell screws until the star image and diffraction rings are perfectly symmetrical. However, I have never felt the need for this in lunar work and am always quite satisfied with the image quality

Fig. 3.19 *Correctly collimated optical components of a Newtonian reflecting telescope*

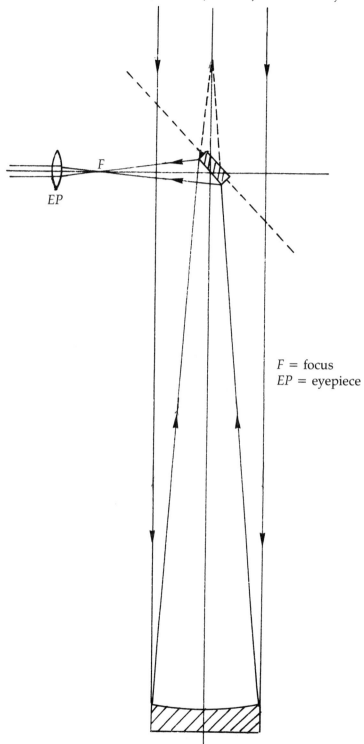

F = focus
EP = eyepiece

Fig. 3.20 View through the eyepiece holder of a properly collimated Newtonian reflecting telescope

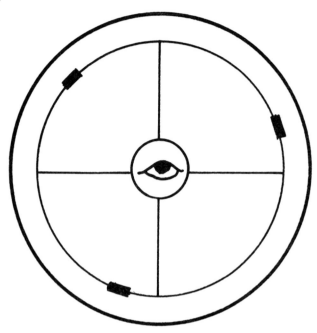

Fig. 3.21 View through the eyepiece holder of an improperly collimated Newtonian reflecting telescope (main mirror out of adjustment)

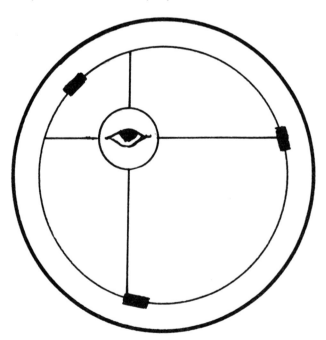

afforded by the accuracy of collimation attained in the method just described.

It is a good idea to get into the habit of checking the collimation of your telescope every time you use it, especially if you 'set up' and 'take down' your telescope each time you use it as I do, because the inevitable slight knocks and jars to the main telescope tube can upset the alignment of the optical train, but not usually seriously.

Setting up and adjusting the equatorial mount

There is no point in purchasing an electrically driven equatorial reflector if you are not going to set it up properly. As explained previously, the polar axis should be parallel to the Earth's axis. When thus set up the polar axis, actuated by the electric motor at its lower (south) end, will rotate in a direction opposite to the rotation of the Earth's axis and at a rate that is opposite to and nearly equal to the apparent motion of the Moon (really the Earth's). Once the telescope is pointed to the Moon it will continue to be so steadily and accurately if all the adjustments are correct. On looking through the eyepiece, the Moon will appear to be stationary. Even if all adjustments are perfect, the Moon will still exhibit a very slow drift in a west to east direction. This is because the electric motor is set to give one complete rotation of the polar axis in a sidereal day (23 hours and 56 minutes). A star would therefore stay in the field indefinitely and exhibit no drift, but because the Moon moves in its own orbit it appears to move slowly eastward against the background of fixed stars as seen from the Earth. This drift due to the Moon's orbital motion is so slow, even when magnified in the telescope, that it is of no consequence to the visual lunar observer.

In order to adjust the polar axis so that it is parallel to the Earth's axis, the following procedure although somewhat crude will yet be accurate enough to suit the requirements of visual lunar observation.

First, slightly loosen the nut that allows the polar axis to be tilted up and down about the big horizontal screw that attaches the equatorial head to the upright pillar. This will be made easier if the telescope tube and counterweights are first removed. With the aid of a protractor, try to tilt the polar axis to make an angle to the ground as nearly as possible equal to your local latitude and tighten up the nut again. This must be so tight that when the counterweights are attached, the polar axis will not slip around on the horizontal screw and tilt out of the correct angle. On the next clear night, check this adjustment by setting up the telescope with the counterweights attached and with the polar axis pointing due north as indicated by the pole star (the higher end of the inclined polar axis should be pointing north). Rotate the telescope tube about the Declination axis so that the open end of the tube is pointing in the general direction of the pole star. Now rotate the polar axis so that the telescope tube is on top of and parallel to the polar axis and the counterweights are underneath. The Declination axis will now be oriented in a vertical plane (Fig. 3.22). If the angle that the polar axis makes with the ground is accurate and if the axis is oriented in a true north–south direction, then on looking through the (preferably low power) eyepiece and focusing, the pole star should be seen in the field of view, albeit perhaps not perfectly central. If not, readjust the mount to point more accurately in a northerly direction. If you still cannot see the pole star, readjust the angle of the polar axis to the ground – it should be possible to arrange things so that you can loosen the nut that holds the polar axis at the correct angle and at the same time look through the telescope eyepiece. If the tube is accurately pointing due north, then slowly raising and lowering the tube about the horizontal screw whose nut you have just loosened should quickly bring the pole star into the field of view. As soon as it is, center it as accurately as possible and tighten up the nut again. When the telescope is subsequently set up so that the polar axis points in the direction of the pole star, the electric motor started and the telescope then pointed to the Moon, the adjustments made will be found sufficiently accurate to keep the Moon in the field of view long enough to permit several minutes of uninterrupted observation. Apart from the Moon's orbital motion against the background of the stars there will almost certainly be a slow drift of the Moon's image in the field due to the inevitable slight

Fig. 3.22 Tube of equatorially mounted reflecting telescope pointing to the north celestial pole

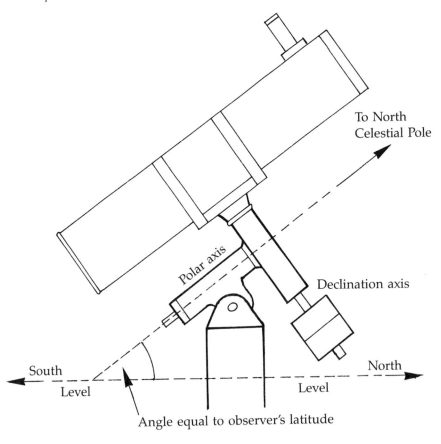

inaccuracies in the angular adjustment and north–south alignment of the polar axis. An occasional slight movement of the polar axis (using the Declination axis as a convenient lever) or a touch of the Declination slow motion control will quickly bring the Moon back to where you want it.

Balancing the telescope

After attaching the counterweights, it is usually advisable, provisionally, to place the telescope tube in its cradle at a little way past the mid point towards the mirror end. Do this with the tube in the horizontal position. Unlock the Declination axis. If the tube is balanced in its cradle, it will stay still and not tilt or rotate about the Declination axis. Remember that the mirror is quite heavy so that proper balance will only be achieved if this is allowed for by placing the tube in the cradle as mentioned a bit further towards the mirror end away from the middle point.

Having pointed the telescope to the Moon, slide the counterweights so that the telescope does not tend to swing around on the polar axis under its own weight. It will do so if the weights are too far up on their end of the Declination axis. If the weights are too far down, they will pull the telescope around in the opposite direction.

Assume now that everything at last is roughly balanced and you have the telescope pointed at the Moon and the latter focused in the eyepiece. You will still probably find, to your annoyance, that the Moon drifts fairly rapidly in the field of view. If so, then obviously the telescope is not following the Moon as it should. First check that the motor drive is running by placing your ear close to the box containing the drive mechanism at the lower end of the polar axis. If you can hear a slight whirring or humming sound, then all is well, as it usually will be if your wiring and electrical contacts are respectively unbroken and clean; do not forget to switch on the current if your electrical outlet is controlled by a switch! Assuming that the motor is running, the trouble is due to the fact that the telescope is still not quite accurately balanced about the polar axis by the counterweights. The counterweights are probably too far down the Declination axis so that the drive is not quite strong enough to overcome their moment about the polar axis. Shift the weights a bit further up the Declination axis, lock them in position and look through the eyepiece again. If the Moon's image holds still, then everything is now perfectly balanced and you can look forward to observing the Moon fairly uninterruptedly except for an occasional touch of the Declination slow motion control or the polar axis, which you can do by using the counterweight shaft (Declination axis) as a lever. (In case you decide to look at some other celestial object in a different part of the sky, do not assume that the telescope will be perfectly balanced in its new position. The balancing procedure will have to be gone through again. Some telescopes have small accessory weights that slide up and down on a long metal rod attached to the telescope parallel to it and adjustment of these may be used to overcome slight imbalances of the tube when its angle and direction are changed away from an object for which it has been perfectly balanced.)

During long observing sessions, the telescope tube changes its direction significantly as it follows the Moon so that in the new position it may become slightly out of balance again. You may find, therefore, that after the periodic slight adjustment that you need to make to the polar axis to keep the object you are studying in the center of the field of view, the telescope acquires an annoying tendency to slip back a little so that the formation that you have just re-centred in the field slips over to the edge. This is because the counterweights are now not exactly balancing the telescope in its new position. A very slight readjustment of the counterweights should cure this. Occasionally, you may find that the counterweights are seemingly too heavy; however far up the Declination axis you slide them, the telescope still refuses to be turned on the polar axis by the motor drive. This is sure to happen in certain positions of the telescope. This is remedied by simply removing one of the weights and positioning the other until you achieve a perfect balance, which should now be possible.

All this may sound very tedious but you quickly get the 'feel' of things and after some practice and familiarity with the idiosyncrasies of your own instrument, the proper setting of the counterweights for any position of the telescope can usually be quickly found.

A few final hints may not be out of place here. First, before you start observing, rotate the telescope tube in its cradle so that the eyepiece holder is on the same side as the counterweights and the Declination slow motion control. You can then make slight adjustments during observing sessions without removing your eye from the eyepiece because the controls will be easily within reach. If the eyepiece holder is positioned on the opposite side of the telescope tube, you will probably not be able to reach the controls without taking your eye away from the eyepiece holder.

Next, if you are observing the Moon when it is high in the sky and near to the meridian, the telescope tube will make quite a steep angle to the ground and the lower (mirror) end may clear the ground by only a few inches. Suppose that you have arranged the telescope tube so as to be on the west side of the polar axis. As the Moon passes the meridian the bottom of the telescope tube swings lower and lower and may eventually touch the ground and the telescope will no longer follow the Moon. This can be prevented by having the telescope tube on the east side of the polar axis. In this position the telescope tube is higher up on the mount and the bottom will be moving away from the ground instead of towards it. If you still have trouble, you can try shifting the tube further up in its cradle. However, if this is overdone, the upper part of the telescope will be top heavy about the Declination axis and you may have to screw the Declination locking device very tightly to stop the top of the telescope tube from swinging downwards under its own weight and hitting the ground. Another way to prevent the mirror end from touching the ground is to raise the tripod feet of the stand on bricks or wood blocks. I have never found this to be really necessary, however.

When making an adjustment about the polar axis, always move the tube in a west to east direction. If you do it the other way around, the object that you have just centered in the field of view of the eyepiece will usually drift for several seconds, just as it does when the motor is off or the telescope is improperly balanced. It will usually stop drifting quite quickly and the telescope will resume following the object, which will now be well away from the centre of the field. This effect has to do with the cogs and gears in the motor drive. If you move the tube correctly in the west to east direction, the cog teeth are all in contact and there will be no lag in the telescope resuming its following of the object. If you turn the telescope tube in an east to west direction, this will disengage some of the motor cog teeth, which then have to move through a tiny distance before they engage again with the cog teeth of the other wheels. This usually takes a few seconds and is the cause of the short stop before the telescope resumes following the object under observation.

Finally, if you have occasion to rotate the telescope from one side to the other of the polar axis, remove the eyepiece and Barlow lens first. The tube turns upside down during this maneuver so that there will be a risk of the eyepiece and Barlow lens falling out if they are not first removed.

Adjusting the viewfinder

The viewfinder or 'finderscope' is the small telescope attached near to the upper end of the main telescope tube so that it is parallel to and pointing in the same direction as the main telescope. The field of view in even a moderate telescope, especially when a high powered eyepiece is used, is so small that the telescope user, even if experienced, will often have difficulty in pointing the telescope quickly and accurately to the Moon. This is even more difficult with a reflector; whereas with a refractor the observer is looking in the same direction as the object being viewed, this is not so with the reflector, where the eyepiece holder is at right angles to the direction in which the telescope is pointing.

The viewfinder telescope has a much wider field of view and so it is much easier to locate objects with it. Usually, an eight-inch reflector will be fitted with a viewfinder, the objective lens of which will be about 50 mm in diameter and the power about 7×. The field of view covered by the viewfinder will be equivalent to several full Moon diameters.

On looking through the eyepiece of the viewfinder, two fine black lines at right angles to each other will be seen intersecting at the middle of the field. These are the 'cross wires' and they are used to center an object exactly in the center of the field of view.

To point the main telescope accurately at the Moon first set up the telescope correctly on its equatorial mount and start the motor drive. Unlock the Declination axis and point the telescope in the general direction of the Moon. Look through the viewfinder and move the main telescope to and fro, up and down until the Moon comes into the field. Position the Moon's disc so that it is exactly on the intersection of the cross wires. Lock the Declination axis. On looking through the main telescope eyepiece you will see the Moon in the field of view, the center of its disc in the center of the field – if the viewfinder has been correctly adjusted. If not, unlock the Declination axis and carefully position the center of the Moon's disc in the center of the field of view. Lock the Declination axis. Then looking through the viewfinder eyepiece, and trying not to jog the main telescope, adjust the direction of the viewfinder with the adjusting screws. This is done by carefully loosening one screw while tightening another until the cross wires are over the center of the Moon's disc. The knack is soon acquired. Look again through the eyepiece of the main telescope to check that the center of the Moon's disc is still located at least approximately in the center of the field of view. The viewfinder will now be sufficiently accurately adjusted for the needs of the visual lunar observer.

On a dark night when the Moon is well past the first quarter phase or is full, its light will be sufficiently bright to cast a clear shadow of the telescope, especially if it is standing on a smooth light-colored surface such as a concrete patio. Instead of squatting down and squinting through the viewfinder while you fumble with the adjustments to get the Moon on the viewfinder cross wires and then locking the Declination axis, you can get the telescope pointing fairly accurately at the Moon by observing the shadow cast by the telescope tube on the ground. After unlocking the Declination axis, point the telescope tube in the general direction of the Moon and then, while observing the shadow it casts, move the telescope slowly until it casts a nearly perfect elliptical shadow. On looking through the eyepiece of the main telescope you should see the brilliant light of the Moon – the actual image will probably be out of focus – or if not, a slight movement of the tube about either axis should bring the Moon into view. Then lock the Declination axis firmly.

Telescope magnification and visibility of lunar detail

The owner of a telescope intended for lunar observation will naturally want to know what is the finest detail on the Moon's surface that the telescope will reveal, and what magnification is needed to make it visible. As was shown earlier, the ability of a telescope to reveal detail is determined by the aperture of the objective lens or mirror; the larger its diameter the finer the detail that will be resolved.

Many years ago Dawes proposed a criterion for expressing the resolving power of a telescope objective. It states that a very close double star is resolved if, under high magnification, the diffraction disc images of the two components overlap by an amount equal to half their diameter. However, this would be rather difficult to see if the seeing conditions were poor or if the observer had poor eyesight or was inexperienced. The angular separation in seconds of arc of the components of a double star that are just resolved by this 'Dawes limit' is considered to be 4.56 divided by the working aperture of the telescope in inches. Thus, a three-inch telescope will just resolve 1.5 seconds of angular separation, a six-inch 0.76 seconds and an eight-inch 0.57 seconds. Another and less rigorous resolution limit is to consider the double star resolved when the edges of the diffraction disc images just touch. The lunar observer will probably find that the Dawes limit and other criteria of telescopic resolution of double stars are difficult to translate into terms of the resolvability of lunar detail. Also, the separation of two close bright diffraction discs on a black background is a rather different situation from resolution of irregular detail of varying degrees of brightness on a light field as is the case with lunar detail. The contrast conditions prevailing at the time of observation also affect the resolvability of delicate lunar detail, quite apart from the aperture of the objective.

In an article by E.A. Whitaker on the limiting lunar detail visible with different telescope apertures, there is a table of the approximate diameters of the smallest craters half filled with shadow and the width of the narrowest dark lines (shadow-filled clefts) that are just visible as such. From this table the following information is taken: for a six-inch telescope the smallest craterlet is 1.5 miles across and the narrowest dark line is 150 yards wide. For an eight-inch telescope the corresponding dimensions are 1.1 mile and 110 yards and for a ten-inch, 0.9 mile and 90 yards.

What is the best magnification to use? Too low a magnification will not render visible all the detail that the objective can resolve. As magnification is increased, more and more detail will be revealed in the image but only up to a point. The minimum that will enable you to use the full resolving power of your telescope is roughly $20D$ where D is the diameter of the objective in inches, i.e., a power in this case of around $160\times$ on an eight-inch telescope. In practice it is desirable to use double this power, which although 'empty magnification' in the sense that no fresh detail will appear, is yet useful in that a more comfortable view of the fine detail is afforded than with only the minimum magnification. Further increase of magnification may be possible on nights of good seeing conditions depending on the optical quality of the objective, apart from its aperture, and the observer's eyesight. Magnifications over $50D$ usually result in 'softening' and dimming of the image and this is definitely 'empty' or useless magnification as there is no resultant advantage. The use of high powers is advantageous when the lenses of the observer's eyes are astigmatic. This quite common defect is caused by unequal curvature of the surface of the eye lens in different parts of the lens. When high powers are used, the exit pupil of the eyepiece is quite small and so a correspondingly smaller part of the total surface of the observer's eye lens will be involved. This will result in minimizing the effects of the astigmatism in the eye.

Generally speaking, I personally find that a magnification of about $30D$ ($240\times$) on my eight-inch reflector is about optimal and this may occasionally be increased to about $40D$ on nights of excellent seeing. Different observers may be more comfortable with somewhat greater or lesser magnification depending on their eyesight and personal taste. These may change as experience increases; for example, I find today that I prefer somewhat higher magnifications than I did when I first started serious lunar observing more than 20 years ago.

Testing the telescope

If you purchased a telescope made by a reputable firm, then you may assume that the objective, mirror or lens, will have been tested for performance before being 'passed'. If you constructed your own instrument then you will already be familiar with tests of optical performance. It is when you buy a telescope second hand, especially if home made, that tests of optical performance should be made before you part with money. The seller of such a telescope should not object to this.

Before testing, the telescope should, of course, be perfectly collimated. Perhaps the best test object of all is a moderately bright star on a clear dark night of perfect steady seeing, i.e., with no undulations in the atmosphere. If all conditions are perfect – the atmosphere, the optical components, collimation and the observer's eyes – then with a power of about 100× the image of a star when in perfect focus should appear as a perfectly clear sharp point of light without any appendages or other irregularities. The focusing tube of the eyepiece should now be slowly moved in and out so that you can examine the intrafocal and extrafocal images of the star. The star should now appear as a disc with a dark spot at the center, which is the silhouette of the diagonal mirror. Surrounding the central bright disc should be seen a series of three of four light concentric rings which should all be perfectly circular and even in intensity both inside and outside the focus. Do not immediately conclude that your telescope is faulty if you see any departure from these appearances because it is not often that the state of the atmosphere permits such perfect results. The testing should be done not just once but on several occasions and in different atmospheric conditions. If something still seems to be amiss after several nights of careful testing, remember that the eyepiece – or even your own eyes – may not be perfect, before blaming the mirror. The bright concentric circles are called diffraction rings.

Testing for correction of spherical aberration

While examining the intrafocal and extrafocal star images, note the brightness of the diffraction rings relative to each other. If, in the extrafocal setting, the outermost ring appears bright and wide, or if, in the intrafocal setting, faint and diffuse, then this is indicative of over-correction of spherical aberration, i.e., the mirror is too 'deep' and is more hyperboloidal than paraboloidal. If the opposite appearances are seen, then under-correction is indicated and the mirror is too 'shallow' and is nearer to a sphere than a paraboloid. Before testing, the mirror must have reached temperature equilibrium because rapid temperature changes can give appearances simulating spherical errors.

Testing for astigmatism

If the extrafocal star image and bright rings appear to be elliptical instead of circular, then astigmatism is to blame and this could be due to a faulty mirror or to your own eyes. While looking through the eyepiece, rotate your head one way then the other. If the elliptical star image also appears to rotate in the same direction as your head, then the astigmatism is in your own eyes. Next, try rotating the telescope tube while keeping your head still. If this causes an apparent rotation of the elliptical star image, then the fault is in the telescope.

In the February 1951 issue of *Sky and Telescope*, R.R. LaPelle describes an alternative observational test for astigmatic corrections which must be done on dark Moonless nights of good seeing and transparency. Using a power of about 100× the well-known 'double double' star Epsilon Lyrae should be studied. Apertures of four inches and over split the two pairs easily and this should be possible at one focal setting. An astigmatic mirror will need to be slightly refocused for each pair and both will be split at one focal setting. This effect might also be caused by a poor diagonal mirror, distortion of the primary due to its being fixed too firmly in its cell or sometimes by poor optical alignment or an indifferent eyepiece. These should therefore be checked before blaming the mirror. If you are certain that the mirror is at fault, it should be returned to the manufacturer for exchange or correction of the defect.

Tests of resolution

The Dawes limit criterion for telescopic resolution has already been defined. In selecting a double star for this test, i.e., one that is just resolved by a given aperture in accordance with this criterion, the components should preferably be yellow in color and about equally bright. If one is much brighter than the other, then its diffraction effects might swamp the image of the other fainter component. Tables of double stars and their angular separations can be found in many textbooks of practical astronomy. Blue double stars are easier to resolve than yellow doubles and red doubles are most difficult.

The following double stars are suggested as tests of resolution by R.R. LaPelle. For a six-inch, Zeta Herculis, in which the components are separated by one second of arc. The magnitudes are 3.0 and 6.5. A power of 200× to 250× is recommended. Also, Eta Orionis, in which the magnitudes are 4.0 and 5.0 and separation again is one second of arc. For apertures of ten to twelve inches, the blue component of the wide yellow and blue double star Gamma Andromedae, which is itself a double. A power of 250× to 300× should split the blue component.

Now that the telescope is properly collimated, set up, balanced and tested, you can commence observing the Moon at the earliest opportunity.

Further Reading

Books

Amateur Telescope Making. Books 1, 2 and 3. Ingalls, Albert A.G. (ed.). Scientific American, Inc., United States of America (1949).
Amateur Astronomer's Handbook. 3rd Edition. Sidgwick, J.B. Faber and Faber, London, England (1971).
The Amateur Astronomer. Moore, P.A. Lutterworth Press (1957).

Astronomical Telescopes and Observatories for Amateurs. Moore, P.A. David and Charles (1973).

Telescope Making for Beginners. Worville, R. Kahn and Averill (1974).

Make Your Own Telescope. Spry, R. Sidgwick and Jackson (1978).

Telescopes for Skygazing. Paul, H.E. Chilton Books, Philadelphia, Pennsylvania, United States of America (1965).

The Amateur Astronomer and His Telescope. New revised edition. Roth, G.D. Faber and Faber, London England (1972)

How to Make a Telescope. Texereau, J. Interscience (1957).

4

The Moon from New to Full

Before commencing serious lunar observation and research the beginner should first become familiar with at least the major formations on the Moon's surface and the phases of the Moon when individual formations can be well seen. To do this we study the Moon each night from new to full, noting the different surface features coming into view as the morning (sunrise) terminator creeps from west to east over the lunar surface.

In the following descriptions we imagine ourselves blessed with cloudless nights and perfect seeing conditions. Each phase of the waxing Moon will be noted as will the formations that become visible on each successive evening. The descriptions of the Moon's appearance each day from new to full are given on the assumption of mean libration. The age of the Moon expressed in days, i.e., the number of days that have elapsed since new Moon, is only a rough indication of the position of the terminator relative to any given surface formation; the actual visibility and appearance of a given lunar formation at a particular phase may not therefore be exactly as stated.

When the Moon is one or two days old, i.e., one or two days after new, the crescent phase is so thin, the Moon so close to the sun and low in the sky after sunset that not much can be seen in the telescope. It is best to wait until about the third day after new Moon.

Third day

The silvery cresent of the three-day-old Moon hanging in the dark sky over the western horizon after sunset on a clear winter or spring evening is a sight of great beauty. Usually the unilluminated part of the disc can be seen faintly lit by the 'Earthshine'. This is the effect charmingly called 'the old Moon in the new Moon's arms'. When viewed in the telescope under low power the Moon will now really look like a great ball hanging in space lit by the light of the sun. Sometimes the Earthshine is so bright that the outlines of the dark maria can be dimly seen. Some of the bright craters like Aristarchus can often be distinctly made out. In fact, Aristarchus and other bright spots have at times been so bright on the dark part of the Moon that the great English astronomer Sir William Herschel (discoverer of the planet Uranus) once thought that he was witnessing volcanic eruptions on our satellite!

In the telescopic view of the three-day crescent the eye is immediately drawn to a dusky oval marking near the terminator, a little more than half way down the crescent. This is the Mare Crisium, the first of the larger maria that become visible in the waxing Moon. It appears roughly elliptical in shape, the longer axis being in the north–south direction. Actually the true shape is elliptical with the longer axis in the east–west direction as may be verified by examining a lunar globe. Since the Mare Crisium lies so close to the limb as seen from the Earth, it is seen greatly foreshortened so that it appears elongated in a north–south direction. The Mare Crisium is about 281 miles from north to south and 355 miles from east to west. When viewed under high power the surface of the mare looks smooth and

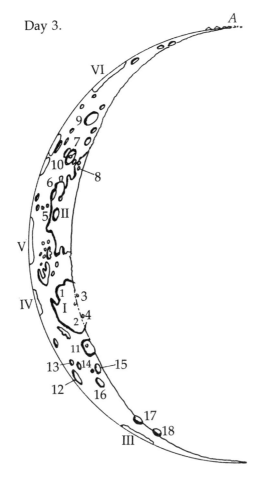

Day 3.

Day 3. Key to formations

1. Cape Agarum
2. Picard
3. Yerkes
4. Peirce
5. Langrenus
6. Vendelinus
7. Petavius
8. Wrottesley
9. Furnerius
10. Palitzsch
11. Cleomedes
12. Gauss
13. Hahn
14. Berosus
15. Geminus
16. Messala
17. Endymion
18. De La Rue

I. Mare Crisium
II. Mare Foecunditatis
III. Mare Humboldtianum
IV. Mare Marginus
V. Mare Smythii
VI. Mare Australe

A. Leibnitz Mountains

Fig. 4.1 *The Southern cusp of the three-day-old moon. April 12, 1986. 20.35–21.05 EST. F.W. Price. Eight-inch reflector*

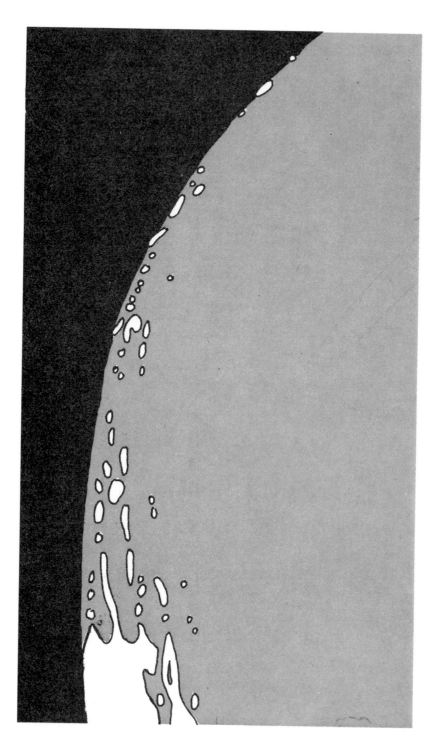

grey. It is bordered on all sides by lofty mountains, those on the west casting dark shadows on the smooth surface upon which some low ridges can be seen running in a north to south direction. On the west, a large 'cape' or promontory protrudes onto the mare. This is called the Cape Agarum. A few smallish shadow-filled craters will also be seen on the surface of the mare, the most prominent being Picard and Peirce. To the west of Picard is a curious whitish patch which is said to be 'variable', while a little to the east and close to the eastern 'shore' is the somewhat sickle-shaped formation Yerkes. The mountain border of the mare east of Yerkes is broken by a gap enclosed by two promontories, Lavinium on the south and Olivium on the north.

To the south of the Mare Crisium ('above' in the inverted telescopic view) is a great chain of four large walled plains, all lying exactly on the meridian of longitude that bisects the Mare Crisium. These are the formations known as Langrenus, Vendelinus, Petavius and Furnerius. The northernmost of the four is Langrenus, which is about 90 miles in length from north to south. At one point on the east the terraced wall rises to a height of nearly 10 000 feet. There is a brilliant central mountain. Next is Vendelinus, a grand formation over 100 miles from north to south. Its walls are intruded upon by two lesser rings on the north-west and north. The one on the north-west, formerly known as Vendelinus C and now named Lamé, is 45 miles in diameter and looks like a smaller version of Langrenus. Webb mentions that a very dark speck can be seen within Vendelinus in the full Moon.

South of Vendelinus is one of the most splendid objects on the Moon, the great ring plain Petavius, about 100 miles from north to south. It has been aptly described by H.P. Wilkins as 'magnificent'. Its oval shape is due to foreshortening owing to its closeness to the lunar limb. Indeed, all features close to the west and east limbs and polar regions are seen in greatly foreshortened perspective so that circular objects are drawn out into elongated ellipses. Petavius has a complex wall which itself consists of several ring-like walls, one within the other, a convex floor and a massive central mountain complex with several peaks. Extending from this to the south-east wall is a prominent shadow-filled valley which is visible even in very small telescopes. This is one of the most prominent of the rills or clefts that abound on the lunar surface. Numerous ridges radiate outwards from the walls of Petavius. If the light is just right, the rim of the crater Wrottesley, which lies against the outside of the east wall of Petavius, may be seen gleaming brightly as the Sun's rays catch it while the interior of the crater is filled with black shadow. Close against the outer west wall of Petavius is the elongated trough-like structure Palitzsch, which P.A. Moore has shown to be actually a chain of five coalesced craters.

One is almost reluctant to leave Petavius but we must move on. Continuing southwards we come to the fourth member of this meridional chain, Furnerius, which is about 80 miles in extreme length. Its terraced walls rise to 11 500 feet in the north and its interior contains much interesting detail. According to H.P. Wilkins and P.A. Moore, there are fourteen large craters and several craterlets on the interior of Furnerius. The most prominent

Fig. 4.2 Petavius and Wrottesley (redrawn after W. Goodacre) (F.W. Price)

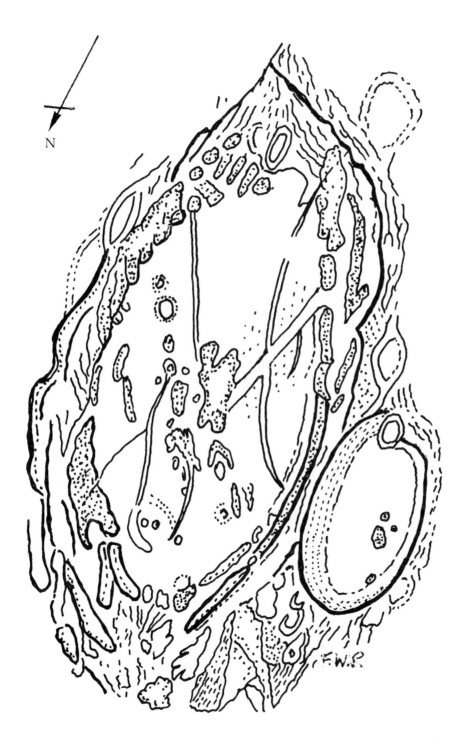

N

object on the floor is the crater B, which is bright and has a central hill. Furnerius appears to have no central mountain.

The three craters Langrenus, Vendelinus and Petavius stand on the west border of a dusky greyish area to the east which is part of the Mare Foecunditatis, the rest of it being in darkness beyond the terminator.

As we continue sweeping southwards towards the southern cusp of the crescent (i.e., the top cusp as seen in the inverted telescopic view) we encounter rugged terrain of mountain peaks and hillocks all casting black shadows and everywhere the lunar surface is pockmarked with innumerable craters, large and small. Beyond the southern cusp and continuing the curve of the crescent will be seen brilliant star-like points of light glittering against the dark sky. These are the peaks of the lofty Leibnitz mountains just catching the first rays of the Sun. One of these peaks is so high that our Mount Everest would be dwarfed beside it.

If now we move the telescope to bring the north 'shore' of the Mare Crisium into view, a large somewhat oblong enclosure will be seen a little way below the Mare. This is the formation called Cleomedes. On its interior are some small hills and craters and many cracks. Cleomedes is about 78 miles across.

North of Cleomedes is the splendid crater Geminus, 55 miles in diameter, with much-terraced inner walls that gently slope to the interior. There is a conspicuous but small central mountain. Within Geminus are two fine but easily visible clefts. T.G. Elger remarks on the warm sepia color of the Geminus region when viewed under a low Sun.

Close to the limb in the general area is a large structure seen foreshortened into an elongated ellipse. This is Gauss, which is actually circular in outline. It is about 110 miles in diameter. A fine mountain chain traverses the floor from north to south. Close to Gauss and in the direction of Cleomedes are two other craters, Hahn and Berosus.

A little further north is another somewhat oblong formation, Messala, which is about 72 miles from end to end. Still sweeping northwards through rugged terrain, the crescent begins to narrow as the northern cusp is approached. The large walled plain Endymion now comes into view. It is 78 miles across and has a dark floor with little detail apart from a few pits. When the light is strong some dark patches can be seen on the floor of Endymion which seem to increase in size and to spread out from night to night.

North of Endymion is the large formation De La Rue, which is not much more than an enclosure bordered by ridges and low irregular walls. There are some craters and ridges on its interior.

As the northern cusp of the crescent swings into the telescopic field, no star-like points of light comparable to those at the southern cusp are seen. Around the pole are some mountains which though not as high as the Leibnitz are yet high enough to be always in sunlight (except during an eclipse) and so they are called the Mountains of Eternal Light.

Before leaving the three-day crescent Moon, there are four maria that may have escaped notice, especially if the libration is unfavorable, because they lie right on the limb. These are the Mare Humboldtianum (close to

Endymion), Mare Marginis (near Mare Crisium), Mare Smythii (on the west limb roughly between Langrenus and Mare Crisium) and Mare Australe (to the north of Furnerius).

Fourth day

The first thing noticed is that the terminator has advanced quite a way eastwards from the position it occupied on the previous evening and the crescent looks thicker. The Mare Crisium and 'magnificent Petavius' are now almost devoid of shadow.

Commencing exploration towards the south, there will immediately be seen on the terminator the huge formation called Janssen with the Sun just rising on it. Janssen is decidedly hexagonal in shape and is over 100 miles across. Perhaps the most interesting feature on its rugged interior is a great branching cleft. One end is broad and deep but it gets narrower towards the other end where it suddenly divides into three branches and curves around towards the south crater wall. A group of four other craters intrude upon Janssen; these are Brenner, Metius, Fabricius and Fabricius A. To the

Day 4.

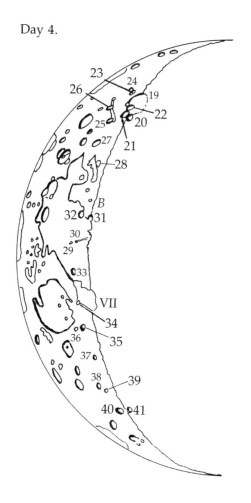

Day 4. Key to formations

19. Janssen
20. Brenner
21. Metius
22. Fabricius
23. Steinheil
24. Watt
25. Rheita
26. Rheita Valley
27. Reichenbach
28. Santbech
29. Messier
30. Pickering
31. Gutenberg
32. Goclenius
33. Taruntius
34. Proclus
35. Macrobius
36. Tisserand
37. Newcomb
38. Franklin
39. Cepheus
40. Atlas
41. Hercules

VIII. Palus Somnii

B. Pyrenee Mountains

south-west on the outside of Janssen is the close crater pair, Steinheil and Watt.

Near to Janssen there is what looks like a huge gouge mark in the Moon's surface. This is the Rheita Valley, named after the large crater at its northern end. The Rheita Valley is really a crater chain.

Further south amidst rugged terrain are two large craters that stand out from the other myriad smaller formations. These are Reichenbach, about 30 miles in diameter and Santbech, about 44 miles in diameter. The walls of Santbech are cut through with valleys on the north and south-east and there is a hill not quite in the center of the floor. Interestingly, many years ago, P.A. Moore observing with only a three-inch refractor discovered a valley-like feature close against the outside of the west wall of Santbech that was not shown on any of the lunar maps of the time.

Fig. 4.3 Proclus. April 28, 1982. Colong. 337.4°. F.W. Price. Eight-inch reflector

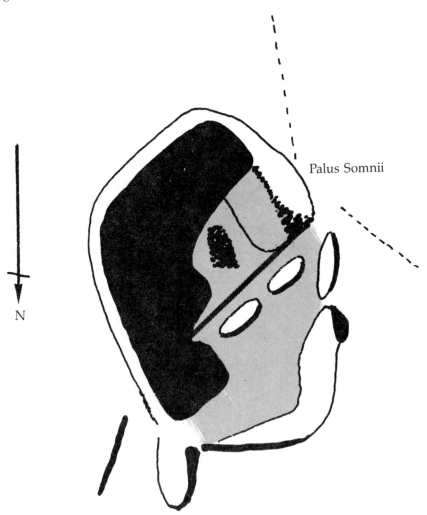

Palus Somnii

N

The westward movement of the terminator since the previous evening has brought the Mare Foecunditatis into full view. An eye-catching feature on its surface is the close crater pair Messier and Pickering and the curious light double streak that extends easterly from Pickering across the mare surface. This double streak has been called the 'comet tail' because of its resemblance to a comet. Messier and Pickering may well be called the 'odd couple' because they seem to change in shape and size from night to night.

There most certainly cannot be any real physical changes in these formations and the peculiar appearances must be due to tricks of the lighting and peculiarities of the structure of the two craters.

Further south on the east border of the Mare Foecunditatis is the somewhat pear-shaped formation Gutenberg; it is traversed by a cleft running south-west to north-east which continues for some distance outside the formation. To the west are three more clefts that originate in the crater Goclenius. Gutenberg appears to be the focus of a system of roughly parallel clefts to the north which is clearly part of a great line of crustal weakness

Fig. 4.4 Macrobius. April 28, 1982. Colong. 337.2°. F.W. Price. Eight-inch reflector

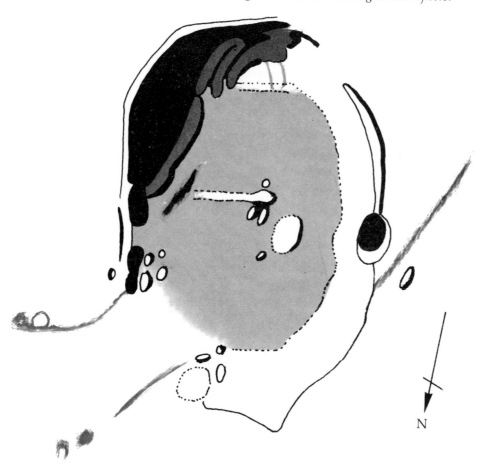

N

reaching all the way from Goclenius to Hyginus. The latter crater is, however, still out of sight on the dark part of the Moon. Extending from the
south and east walls of Gutenberg are the Pyrenee Mountains.

In the northern Mare Foecunditatis is the isolated crater Taruntius, 44
miles in diameter. Even though it has low walls it is a prominent object at
sunrise. On its floor is an inner ring concentric with the outer rampart.

To the north of Taruntius is a curious lozenge-shaped area covered with
short low ridges. It has a peculiar light brown color which distinguishes it
from the general color of the lunar surface. This formation is the Palus
Somnii. Standing right on the edge of the Palus facing Mare Crisium is a
small brilliant crater called Proclus. From it radiate several delicate ill-defined
light streaks that spread fan-wise over the Mare Crisium. These streaks are
best seen under a high Sun and are one of the smaller of several bright
'ray systems' that are found on the Moon's surface. Proclus is one of the
brightest spots on the Moon. There is some debate as to whether there is
a central elevation in Proclus. I have seen one or two short ridges near the
center of the floor but would not consider any of them as central elevations
in the usual sense of the term. Running across the floor I have seen a dark
line, visible just when the sunrise shadows have receded. It requires very
steady seeing. Nothing like this is mentioned in the description of Proclus
in Neison, Elger, Webb or Wilkins and Moore but Goodacre says that Schmidt
shows a longitudinal cleft on the floor which is probably the dark line.

North of the Palus Somnii is another somewhat isolated crater, Macrobius
42 miles in diameter. There is a smaller companion ring to the west called
Tisserand. Macrobius has a central mountain but it is not prominent. Wilkins
and Moore mention that it has two peaks; I have seen it as consisting of
four separate components with my eight-inch reflector. I have also seen a
low dome-like object to the north-east of the central mountain and to the
north a tiny hill. Surprisingly, neither of these two features is mentioned
specifically by Elger, Goodacre, Neison, Webb or Wilkins and Moore. Sometimes the floor of Macrobius is very dark.

Fig. 4.5 Messier and Pickering. Colong. 328.7° (from a photograph) (F.W. Price)

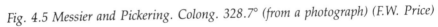

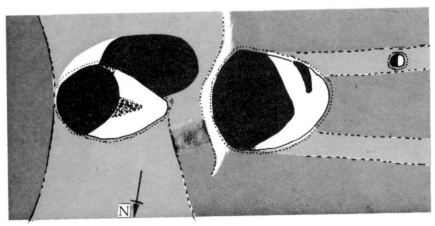

Passing rapidly through rugged country north of Macrobius we pass the craters Newcomb and the Franklin–Cephus pair and just to the north of these is another splendid pair, Atlas and Hercules, a wonderful sight under low illumination. Atlas, 55 miles in diameter, is somewhat polygonal in outline and has but a small central mountain. There are several clefts on its floor and two dusky patches, one to the north with a circular outline and another in the south. They are most prominent under a high Sun and seem to vary in tint, darkness and brightness in a most curious manner if studied over an extended period of time. Hercules, the companion to Atlas, has a deep crater on its interior with a sharp rim.

Here, we are coming close to the northern cusp of the crescent, a rugged area abounding with ring structures. With this final observation we conclude our survey of the four-day-old Moon.

Fifth day

The steady eastward movement of the terminator has now brought two more of the maria into view, the Mare Nectaris and the Mare Tranquillitatis. About half of the latter is visible at this phase.

Day 5.

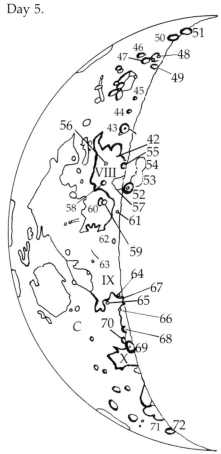

Day 5. Key to formations

42. Fracastorius	61. Torricelli
43. Piccolomini	62. Maskelyne
44. Stiborius	63. Cauchy
45. Wöhler	64. Vitruvius
46. Rosenberger	65. Maraldi
47. Vlacq	66. Littrow
48. Hommel	67. Mt Argaeus
49. Pitiscus	68. Le Monnier
50. Mutus	69. Posidonius
51. Manzinius	70. Chacornac
52. Theophilus	71. Gärtner
53. Cyrillus	72. Arnold
54. Catharina	
55. Beaumont	VIII. Mare Nectaris
56. Rosse	IX. Mare Tranquillitatis
57. Mädler	X. Lacus Somniorum
58. Daguerre	
59. Isidorus	C. Taurus Mountains
60. Capella	

The Mare Nectaris is roughly square in outline and is about 180 miles across. The large horseshoe-shaped formation on its south 'shore' is Fracastorius, about 60 miles across, the north wall of which seems to have been destroyed by a flood of molten lava fom the Mare Nectaris at some early epoch of the Moon's early history, which has spread right into the interior of Fracastorius. The floor of this great enclosure has much interesting detail and will repay careful scrutiny. Delineations of the floor detail by different observers are not in exact agreement. Near the center is a curious object made up of four light spots in a square configuration and in the center of these is a craterlet. T.E. Elger mentions that these objects undergo 'notable changes of aspect' at different phases.

Moving southwards from Fracastorius we come to a fire crater, Piccolomini, 56 miles in diameter, with complex terraced walls. It has a bright central mountain seemingly sliced in two by a ravine. Apart from this there does not seem to be much more floor detail.

To the south of Piccolomini are the smaller craters Stiborius and Wöhler. The four formations Fracastorius, Piccolomini, Stiborius and Wöhler form a chain of craters that decrease in size from north to south. For this reason it is called a decremental chain. There are other examples on the lunar surface.

At this phase we have a better view of Janssen and its floor detail now that the Sun is a little higher.

South of Janssen is a magnificent cluster of large walled formations – Rosenberger, Vlacq, Hommel and Pitiscus. Hommel is a very irregular formation about 75 miles across; on every side except the west it is encroached upon by ring plains and lesser depressions. Of these Hommel A on the north is most notable with very brilliant walls, a conspicuous central mountain, a crater on its floor and other details. Vlacq, adjoining Hommel on the north-west, is about 56 miles in diameter and has terraced walls. The south wall is broken by a fine crater and on the floor are many small craters, short low ridges and a central mountain. The west and south-west walls of Vlacq impinge upon Rosenberger, 50 miles in diameter, which has a dark floor and a central hill. The inner slopes of the low walls are very rugged. Just to the north-east of Hommel lies the 52-mile formation Pitiscus, the most regular of the Vlacq group of craters. The wall is continuous except on the east where it is interrupted by a crater. On the north part of the floor center is a crater. Towards the southern cusp of the crescent are the large formations Mutus, 50 miles across with lofty terraced walls, and Manzinius, with wide terraced walls and, according to T. G. Elger, a small central peak visible only in large telescopes. Returning to the Mare Nectaris, we notice on the east close to the terminator a prominent ridge running from the great crater Theophilus to the crater Beaumont. Theophilus and the two craters Cyrillus and Catharina form a splendid trio to the east of the Mare Nectaris. This evening only their rims are catching the first rays of the Sun so we must wait until tomorrow evening for a better view of them.

The largest crater on the surface of Mare Nectaris is Rosse, diameter 10 miles. A whitish streak runs from it in a north-west direction. In the north-

Fig. 4.6 Posidonius. April 4, 1976. Colong. 334.6°. F.W. Price. Eight-inch reflector

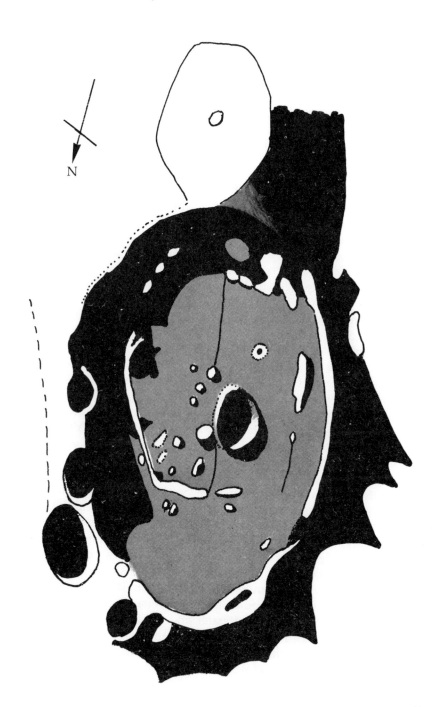

east 'corner' of the Mare is the crater Mädler, 20 miles in diameter and to the west is the somewhat larger Daguerre, the walls of which are merely low ridges.

North of Daguerre in light-colored upland terrain is a crater pair, Isidorus and Capella. The rampart of Isidorus rises as a peak on the west to a height of more than 13 000 feet above its floor. The floor contains no detail except for a small bright crater on the east and another slightly smaller one next to it on the north. Capella, the western companion of Isidorus, is a little larger with a diameter of about 30 miles. It slightly overlaps Isidorus. Its walls are terraced and are cut through by rill valleys on the south-east. The north-west rampart is cut through by a valley with a curious keyhole shape. Capella has a fine central mountain.

North of Mädler and lying on the dark marial surface connecting the Mare Nectaris with Mare Tranquillitatis is an egg-shaped crater, Torricelli, which consists of two unequal coalesced craters. The larger (western) ring is about 10 miles across. To the east is a delicate rill running in a north–south direction and another on the north running in an east–west direction.

Moving further north we come to the Mare Tranquillitatis. On the south part of the Mare is the crater Maskelyne, 19 miles in diameter, about the only sizeable crater visible on the Mare at this phase. It has a distinct central mountain. Some distance to the south-west is the much smaller crater Cauchy, only 8 miles across. The main interest in this area is not Cauchy itself but the two long clefts that run in an easterly direction, one on each side of Cauchy.

Sweeping northwards along the terminator we encounter some light-colored terrain upon which the craters Vitruvius, Maraldi and Littrow are situated. Here also is Mount Argaeus, which is a striking and beautiful spectacle at sunrise. It attains a height of over 8000 feet above the Mare surface. To the north of Mount Argaeus the most westerly part of the Mare Serenitatis is now visible, Here is the incomplete ring of Le Monnier, appearing like a small bay indenting the 'shore' of the mare. To the west of Le Monnier is the light-colored upland area known as the Taurus Mountains.

A little further to the north of Le Monnier on the Mare border the eye is immediately attracted to a splendid crater named Posidonius and its southerly companion Chacornac. Posidonius, about 62 miles in diameter, is one of the most interesting formations on the Moon. The wall is narrow and at sunrise the east wall casts a long serrated shadow. On the floor are many interesting objects. There is the remnant of what was once a second ring concentric with the main crater wall. The most prominent object is a crater a little to the east of center. To the west of this is a curved group of hillocks. Between these and the central craterlet runs a long cleft, difficult to see in its entirety unless under a low Sun. There is another cleft on the north-east side of the floor and a short transverse cleft on the south part of the floor. Between the central crater and the south-east wall is a tiny craterlet that becomes surrounded with a bright halo under high illumination. In the north part of the floor around the somewhat bent and flattened north end of the inner ring wall are several somewhat ill-defined hillocks.

Posidonius lies on the north shore of one of the smaller marial bodies, the Lacus Somniorum. Further north the terminator is seen to cut through the western extremity of the elongated Mare Frigoris. On the north border of this mare in this region lies the incomplete crater ring Gärtner, the side facing the mare having lost most of its wall.

Not far from Gärtner in a northerly direction is the formation Arnold, somewhat smaller than Gärtner, enclosed by straight parallel walls. The interior has a dark tone and contains a small crater.

Here we are close to the north cusp of the crescent, a rugged area abounding with ring structures. With this final observation we conclude our survey of the five-day-old Moon.

Sixth day

The grandest sight on the six-day-old Moon is undoubtedly the curving group of three great craters to the east of Mare Nectaris – Theophilus, Cyrillus and Catharina, – of which we had a glimpse on the previous evening. The northernmost of the trio, Theophilus, is 65 miles across and its massive walls rise in places 3½ miles above the interior. Below this lofty mountain ring are two more concentric rings, each successively lower going inwards until the floor is reached. Although Theophilus seems to be circular under low telescopic power, higher magnification shows the wall to be composed of roughly linear segments so that it really has a polygonal outline. There is a magnificent bright multiple central mountain, one peak of which rises about 6000 feet above the floor. A curious thing about the central mountain is that its contour seems to change as the lunation progresses. W.H. Pickering considered these appearances to be due to deposit, and later melting, of snow! Observers in the USA have studied these phenomena and find that Pickering's drawings do not now represent the sequence of changes as presently observed.

The walls of Theophilus encroach upon its southerly neighbor Cyrillus, which has a somewhat square outline and is 55 miles across. The complex walls of Cyrillus also exhibit abundant linear segments and within the east wall is a small brilliant crater, Cyrillus A. There is a triple mountain near the center of the floor. Cyrillus is connected to Catherina to the south by a broad valley. Catharina is the largest of the three formations and is over 70 miles in extent in the north–south direction. The most remarkable object on the floor is a curved ridge that forks on the east. The curve of the north prong of the fork is continued by an arc of three craterlets. For all their magnificence we must not linger over Theophilus, Cyrillus and Catharina if we are to complete our survey of the six-day-old Moon.

Extending from the south-east of Cyrillus and curving round in a great arch towards the crater Piccolomini is the Altai scarp, 315 miles long and best seen after full Moon. The highest peak, west of the crater Fermat, rises to 13 000 feet. Undoubtedly, the scarp is the border of what was once an ancient 'sea' the Mare Nectaris being all that remains of it.

Just south-east of the Altai scarp and close to Fermat lies the great walled plain Sacrobosco. It is irregular in shape and about 52 miles in diameter, with broad lofty walls. There are many objects on its interior, including two craters A and B each with a central peak, and another crater C to the north of A.

To the north of the Altai lies a cluster of large walled formations – Rabbi Levi, Zagut, Lindenau and Riccius. The largest is the irregular-shaped Rabbio Levi, about 50 miles in diameter. The wall is difficult to trace except under low light. On the south these are cut by a series of parallel ravines. Zagut, the most easterly of the group, is 45 miles across. There is a crater near the center of its floor and a large ring plain about 20 miles in diameter intrudes upon the floor on the north-west. North-east of Zagut is the crater

Day 6.

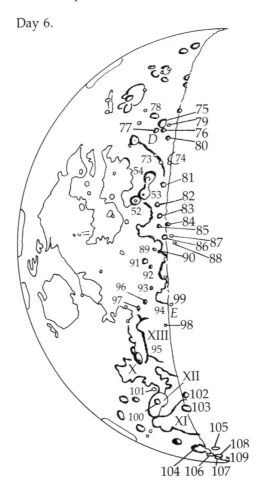

Day 6. Key to formations

73. Fermat
74. Sacrobosco
75. Rabbi Levi
76. Zagut
77. Lindenau
78. Riccius
79. Celsius
80. Wilkins
81. Tacitus
82. Kant
83. Zöllner
84. Taylor
85. Alfraganus
86. Delambre
87. Theon Jr
88. Theon Sr
89. Sabine
90. Ritter
91. Lamont
92. Arago
93. Ross
94. Prom. Acherusia
95. Serpentine Ridge
96. Plinius
97. Dawes
98. Bessel
99. Menelaus
100. Bürg
101. Plana
102. Eudoxus
103. Aristoteles
104. Meton
105. Barrow
106. Scoresby
107. Main
108. Challis
109. Gioja

XI. Mare Frigoris
XII. Lacus Mortis
XIII. Mare Serenitatis

D. Altai Mountains
E. Haemus Mountains

112

Celsius that was named by J. Schmidt and south-west is a formation formerly named Zagut S by W. Goodacre and now called Wilkins after the late H.P. Wilkins, the noted English selenographer. North of Rabbi Levi is Lindenau, about 35 miles in diameter. It has finely terraced walls rising to 1200 feet on the east and there is a group of low hills near the middle of the floor.

Riccius, to the south-west of Rabbi Levi and 50 miles in diameter, has several craters on its floor and a crater on its south wall.

Between the Rabbi Levi group and the southern tip of the crescent is a rugged upland area scattered all over with formations large and small. Returning to Sacrobosco we pause to examine some formations situated in the light upland area to the north and extending to the Mare Tranquillitatis. First there is Tacitus, situated to the north-east of Catharina. It has a polygonal outline and is 25 miles in diameter. The walls are terraced and on the floor is a central crater pit. Southwards are Kant and Zöllner. Kant is a prominent deep ring plain 23 miles in diameter. According to J. Schmidt it has a double rampart. The inner walls are terraced and there is a bright central peak. Zöllner is a larger enclosure; its walls are low but broad with some craters on them.

The crater pair Taylor and Taylor A are to the north-east of Zöllner and to the east of Torricelli. Taylor is elliptical, about 25 miles from north to south, and has a large central mountain. The walls appear continuous except in the north and south where they are broken by craters. Taylor A is 18 miles in diameter. T.G. Elger mentions two ill-defined markings on the floor. Due west of Taylor is Alfraganus, a small crater 12 miles in diameter and the center of one of the small bright ray systems on the Moon. It has a central hill.

The formation Delambre, north of Taylor, is 32 miles in diameter. It has a polygonal outline and terraced walls. There is a low central peak on the interior and on the summit of the north wall is a distinct bright crater. To the east and north-east are the little craters Theon Junior and Theon Senior.

At this phase the Mare Tranquillitatis is now fully visible and we notice immediately the many wrinkle ridges spread over its surface. These are remarkably contorted in places. There are several interesting craters on the mare surface that were hidden beyond the terminator on the previous evening. A close pair, Sabine and Ritter, lie close to the south-east 'shore'. They are closely similar in size, respectively 18 and 19 miles in diameter, both with narrow walls and central hills. The most notable features in this area are the two parallel clefts that run in a westerly direction on the Mare surface. They originate on a mountain arm extending from the south wall of Sabine.

North-west of the Sabine - Ritter pair is the low and imperfect ring of Lamont; to their north is the crater Arago, 18 miles in diameter, in the neighborhood of which, on the east and north, are large but low domes visible in small telescopes. North of Arago is Ross, which is similar in size. To the north again is another 'headland', the Promontorium Acherusia, at the western extremity of the Haemus Mountains, which are presently invisible and in darkness beyond the terminator. It rises to nearly 5000 feet above the Mare Serenitatis, the west half of which is exposed at this phase.

Fig. 4.7 Theophilus, Cyrillus and Catharina (from a photograph) (F.W. Price)

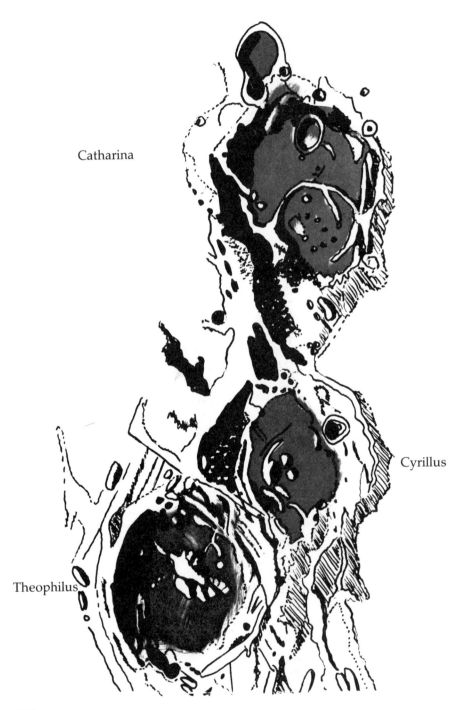

Catharina

Cyrillus

Theophilus

A prominent feature of this part of the surface of Mare Serenitatis is the great Serpentine Ridge, which is well displayed at this phase. It is one of the loftiest and longest of the wrinkle ridges that occur on the surfaces of the lunar maria. Commencing not far from the north-east wall of Posidonius, it winds southwards over the mare surface sending out branches and finally peters out close to the Promontorium Acherusia.

To the south-west of the Promontorium is the fine isolated crater Plinius, 30 miles in diameter, the walls of which are almost circular and finely terraced. There is an object in the center of Plinius that looks curiously different under various illumination conditions. It is variously recorded as a mountain (Schröter), a double mountain (Lohrmann), central craters or low mounds (Mädler) and two central mountains (Neison). Goodacre remarks that the bright central mountain sometimes resembles a double crater. As the Sun's height increases, the crater floor is seen covered with a confusion of light streaks and spots that require a good telescope to be clearly seen. Under high illumination, Plinius looks curiously like a wheel, the spokes of which consist of separate spots according to Martz, Ross and Haas.

To the north of Plinius are three parallel clefts, that closest to Plinius being the broadest; in part it is a crater chain.

In the gap between the Mare Tranquillitatis and the Mare Serenitatis stands the smaller crater Dawes, 14 miles in diameter. According to Goodacre, Dawes has a small central peak, but F.H. Thornton was unable to find it with his 18-inch reflector. The surface of the Mare Serenitatis is dotted all over with craterlets. The largest crater on the mare surface is Bessel, about 12 miles in diameter. It is situated on a light streak that crosses the mare. This streak originates from the 20-mile crater Menelaus on the north 'shore' of mare Serenitatis. South of Posidonius is the crater pair Bürg and Plana, situated on a small marial area, the Lacus Mortis.

Close to the terminator to the east of the Lacus Mortis is the fine crater pair Eudoxus and Aristoteles. Eudoxus, 40 miles in diameter, has lofty walls rising to 1000 feet on the west and they are prominently terraced. There is a circular arrangement of hills on the south part of the floor. On February 20, 1877, Trouvelot saw a delicate thread of light across the south part of the crater interior stretching from border to border. This could have been the top of a wall catching the Sun's light – except that there does not seem to be a real wall there at all when the crater is observed under other illumination angles. Trouvelot saw similar appearances on other parts of the lunar surface and called them *Murs enigmatiques* (enigmatic walls). Aristoteles is a splendid walled formation 60 miles across with a hexagonal outline and terraced walls with peaks, the highest rising to 11 000 feet. There are many hills on the floor and the remains of an old crater ring. A remarkable feature of Aristoteles are the rows of hillocks that radiate outwards from its walls, which look like crowded points of light among the shadows at sunrise.

Crossing northwards over the western extremity of Mare Frigoris brings us to a group of craters close to the north pole. The largest of these, Meton, is one of the largest formations on the moon and is over 100 miles long. It has the appearance of being formed by coalescence of several ring forma-tions. On the light grey interior, which looks very rugged under oblique illumination, is a small mountain near the north rampart. To the east is the formation Barrow with broken walls, 45 miles in diameter. It is a fine telescopic object. To the north are the craters Scoresby, Main, Challis and Gioja, the last of these being very close to the pole. All are seen in very foreshortened perspective.

Seventh day

The Moon is now close to the 'first quarter' or 'half Moon' phase. True dichotomy, when the terminator runs exactly down the center of the lunar disc, occurs about 7½ days after the new Moon. At this phase the formations near the terminator are seen in plan view in the telescope and we really seem to be suspended in space looking down on the Moon's surface. It is a spectacular phase. Glancing over to the western limb, the mountain walls of the Mare Crisium now seem to have disappeared and the mare surface is covered with white spots and streaks. The great craters Petavius and Vendelinus are almost invisible and without relief effects. Many of the other craters will also seem to have vanished unless they are very bright or dark so that they stand out from the general brightness of the surface.

At this phase it is evident that much of the southern hemisphere of the Moon consists of light-colored upland terrain which is crowded with craters of all sizes. Among the craters near the south pole are Schomberger and Simpelius. To the north-east and lying astride the terminator in a meridional alignment are three large craters that decrease in size from south to north: Curtius (50 miles), that has a large dome on its high eastern rampart, Zach (46 miles) with high walls rising to over 13 000 feet, and Lilius (32 miles) with steep finely terraced walls, a fine central mountain and several crater-lets on its floor. To the north-west and south-west respectively of Lilius are the craters Cuvier and Jacobi, forming a north–south meridional pair, respec-tively 50 miles and 41 miles in diameter. Cuvier has terraced walls and on its floor and lower inner slopes are craterlets that appear as bright spots at full Moon. The walls of Jacobi are broken on the south-east by the crater J and on the north by an incomplete ring P. There is a prominent crater on the center of the floor and much other detail.

To the north-east of Cuvier is the curiously shaped formation Heraclitus, which is partly intruded upon on the north by the ring plain Licetus and on the south by an apparently unnamed crater. Heraclitus looks like a short section of a huge car tire mark on the Moon's surface or the bottom of a rowing boat. There is a ridge in the center that partly projects into Licetus. Licetus is an irregular formation about 46 miles across and much depressed below the level of Heraclitus.

To the north of Cuvier is the giant enclosure Maurolycus, one of the grandest of its kind on the Moon's surface. T.G. Elger aptly remarks that the sight of Maurolycus under a low Sun 'presents a spectacle which is not easily effaced from the mind'. It extends 150 miles in the east–west direction according to Elger but Goodacre says 72 miles. Several ring structures are

Day 7.

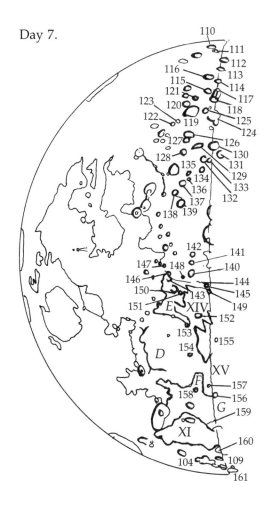

Day 7. Key to formations

110. Schomberger	130. Aliacensis	150. Julius Caesar
111. Simpelius	131. Werner	151. Boscovitch
112. Curtius	132. Apianus	152. Manilius
113. Zach	133. Krusenstern	153. Sulpicius Gallus
114. Lilius	134. Playfair	154. Linné
115. Cuvier	135. Azophi	155. Conon
116. Jacobi	136. Abenezra	156. Cassini
117. Heraclitus	137. Geber	157. Thaetetus
118. Licetus	138. Almanon	158. Calippus
119. Maurolycus	139. Abulfeda	159. Alpine Valley
120. Barocius	140. Agrippa	160. W.C. Bond
121. Clairaut	141. Godin	161. Shackleton
122. Büsching	142. Lade	
123. Buch	143. Ariadaeus cleft	XI. Mare Frigoris
124. Stöfler	144. Hyginus cleft	XIV. Mare Vaporum
125. Faraday	145. Hyginus	XV. Palus Nebularum
126. Gemma Frisius	146. Ariadaeus	
127. Goodacre	147. Dionysius	E. Haemus Mountains
128. Pontanus	148. Silberschlag	F. Appenine Mountains
129. Poisson	149. Mount Schneckenberg	G. Caucasus Mountains

closely crowded together here and give the impression of mutual deformation when the Moon's crust was plastic at some remote time. There are examples of almost every kind of lunar object on the massive ramparts of Maurolycus. A fine rill curves round the outer slope of the west wall a little below its crest and is easily seen in an eight-inch reflector when the east wall is just on the morning terminator. The compound central mountain is lofty and its highest peaks can be seen at sunrise long before the Sun has reached the shadow-filled floor. Beer and Mädler mention that at full Moon Maurolycus is seen to be crossed by 12 diverging bright lines. Close against the south-west wall is Barocius, 50 miles across with a nearly central crater on its floor. The terraced walls of Barocius are broken by the crater B on the north-west, which has a central peak.

Between Barocius and Cuvier is Clairaut, a walled plain 30 miles across that has been deformed by intrusion of two other craters A and B on the south.

Fig. 4.8 Lade. May 4, 1987. Colong. 352.0° app. F.W. Price. Eight-inch reflector

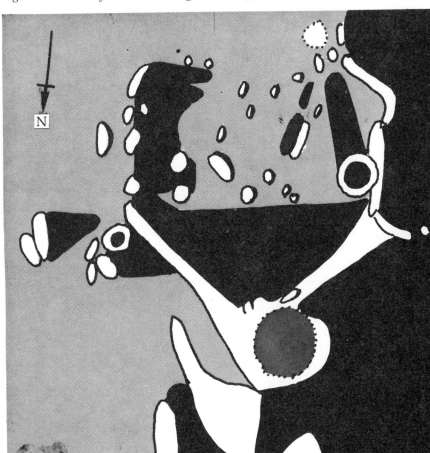

Fig. 4.9 Godin. February 20, 1983. Colong. 10.4°. F.W. Price. Eight-inch reflector

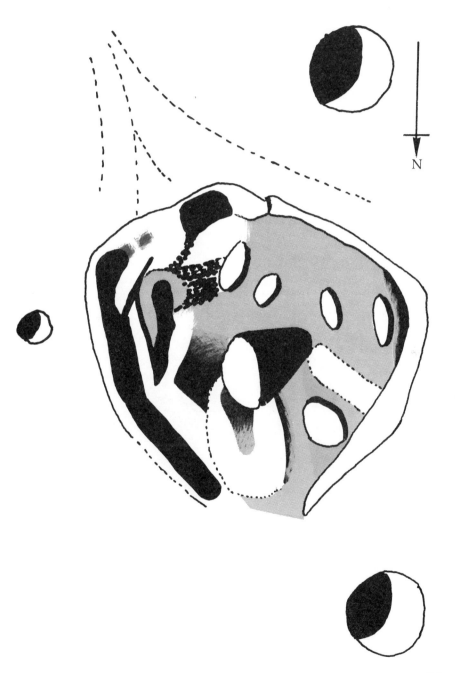

N

North-west from Maurolycus are the close pair of craters Buch and Büsching, respectively 30 miles and 36 miles in diameter. Almost due east from Maurolycus is another large irregular enclosure, Stöfler, which is reminiscent of Maurolycus in size and general features. It is best seen when the morning terminator is slightly east of the west wall. The crater floor seems quite level and is crossed with white streaks from the crater Tycho. They are best seen under a high Sun. The walls of Stöfler are rugged and undamaged in the east. To the south-west Stöfler is encroached upon by the large formation Faraday, about 35 miles across. In its turn Faraday is overlapped by two lesser craters on the south-east and by two still smaller formations on the north-west.

Northwards from Maurolycus is Gemma Frisius, a great walled plain 80 miles from north to south. At one place the wall rises to 14 000 feet above the floor. On the north the fine ring plain Goodacre intrudes upon it. On the floor of Gemma Frisius is a nearly central hill and there is a dark spot on the west. Moving northwards over rugged upland terrain we encounter

Fig. 4.10 Agrippa. April 9, 1984. Colong. 17.4°. F.W. Price. Eight-inch reflector

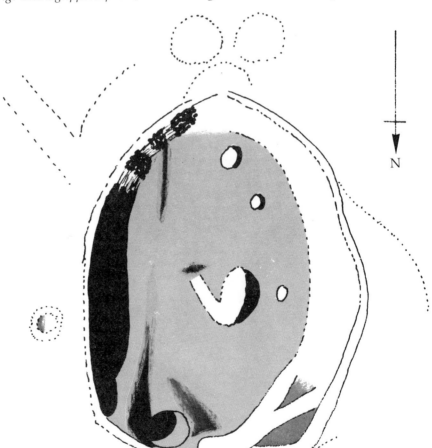

several notable formations. Between Gemma Frisius and Sacrobosco is Pontanus, an irregular formation 28 miles across which has a low broken border, a dark floor and a central mountain.

North-east of Gemma Frisius is the irregular formation Poisson and eastwards from Poisson lies the fine crater pair Aliacensis and Werner. Aliacensis, 52 miles in diameter, has terraced walls that rise to a height of no less than 16 500 feet on the east. The inner west slope has much complex detail which is a fine sight when the illumination angle is suitable. Its companion crater, the 45 mile Werner, closely adjoins Aliacensis on the north-west. On the inner slope of its north-east wall is a bright spot that T.W. Webb considers has faded since Beer and Mädler's time.

To the north-west of Aliacensis and Werner are three craters, Apianus, Krusenstern and Playfair. Apianus, 39 miles across, has a dark grey floor apparently without much detail. Krusenstern is bordered by ridges and Playfair, 28 miles across, is connected to Apianus by a mountain arm.

To the east of Playfair is a large unnamed enclosure 60 miles from north to south. At this phase, when the terminator bisects Aliacensis, the jagged shadows of the east wall of Playfair and the mountain arm stretch across the grey-blue surface of this enclosure and make a fine spectacle when viewed in a large telescope. Proceeding in a north-west direction from Aliacensis and passing beyond Apianus, Krusenstern and Playfair, we come

Fig. 4.11 The Hyginus cleft and Mount Schneckenberg (after V.A. Firsoff) (from Surface of the Moon *Hutchinson, 1961)*

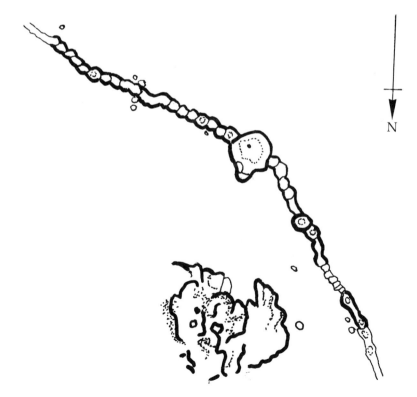

N

to a close crater pair, Azophi and Abenezra. Azophi is 25 miles in diameter and has a rather dusky interior with some light spots. Abenezra is 27 miles in diameter and under morning illumination seems to be cut into two by a curved ridge.

Still moving in a north-west direction brings us to Geber, 25 miles in diameter, with remarkably regular walls. Still continuing in the same direction we come to Almanon and its companion Abulfeda to the north-east. Almanon is about 36 miles across and has an irregular polygonal outline. Running tangentially from the north wall is a remarkable straight crater chain that makes tangential contact with the south wall of Abulfeda. Abulfeda is larger and more massive than Almanon. It has a diameter of 40 miles and its walls are beautifully terraced and much higher than those of Almanon. Wilkins and Moore state that the northern half of the floor of Abulfeda is darker than the southern but my own telescopic observations show that it is the eastern two thirds of the floor that is darker than the western.

Fig. 4.12 Linné. April 20, 1953. Redrawn after P.A. Moore. 33-inch refractor (Meudon) (from Our Moon, *by H.P. Wilkins, Muller, 1954)*

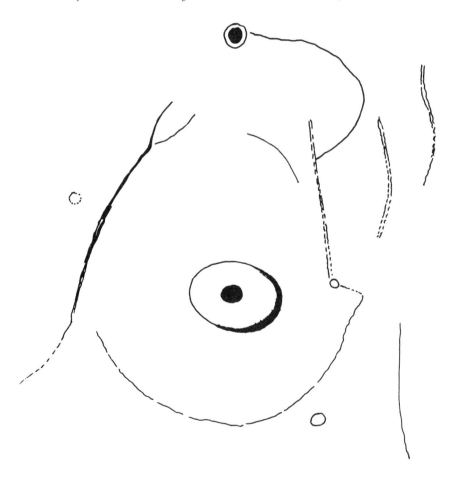

Some distance to the north of Abulfeda and slightly to the east is the fine meridional crater pair Godin and Agrippa. Godin, 27 miles across, is somewhat heart-shaped with terraced inner slopes and a central hill. On the south is a valley shaped like a trumpet, which is a notable feature at sunrise. Slightly larger than Godin and due north is Agrippa (28 miles), shaped rather like a Gothic arch, with an almost straight north wall. There is a large central mountain. To the south of Godin is the incomplete hexagonal formation Lade.

To the north-west and north-east of Agrippa are two of the best known clefts on the Moon, respectively the Ariadaeus cleft and the Hyginus cleft. The Ariadaeus cleft starts close to the 12-mile crater Ariadaeus (just north of Dionysius) after which it is named. It proceeds almost straight in a roughly north-east direction to a point due north of Agrippa, passing close to the crater Silberschlag on its south side. It is over 150 miles long and can be seen in small telescopes but is difficult to find at full Moon.

To the south-east of where the Ariadaeus cleft terminates, the Hyginus cleft begins. This cleft is really a crater chain and I have seen it as such in a three-inch refractor. It runs in a direction parallel to the Ariadaeus cleft until it connects with Hyginus and then continues on its way on the other (east) side of Hyginus but changes direction more towards the north. South of Hyginus is the curious spiral mountain called Schneckenberg.

Northwards from the Ariadaeus cleft is the large imperfect formation Julius Caesar. The floor gradually darkens towards the north where it becomes one of the darkest areas on the Moon's surface. Scattered all over the interior are low hills and craterlets. To the east is another incomplete smaller formation, Boscovitch, whose dark floor is said by Wilkins and Moore to be subject to variations in tint. Like Julius Caesar, Boscovitch is one of the darkest patches on the lunar surface. In a direction slightly north from Boscovitch is the crater Manilius, situated in a marial area that is a part of the Mare Vaporum (most of which has appeared since the previous evening). It has a diameter of 25 miles. Under a high Sun it is very bright and is the center of a faint minor ray system. Manilius is one example among many on the Moon's surface of a 'banded' crater. In this case two dusky bands extend to the east wall from the center.

Proceeding northwards over the Haemus mountains we pass the little crater Sulpicius Gallus to the west as we enter the west part of the Mare Serenitatis, on the surface of which is the celebrated crater Linné that appears as a whitish spot under high light and as a craterlet under low angle illumination. It is one of the most written-about objects on the Moon as some kind of a physical change is supposed to have occured there. Before 1866 it was described as a deep crater five or six miles in diameter and easily visible under all angles of illumination, but it has not looked like this since that time. Considerable controversy has raged since then over the reality of the supposed change which has occurred here, namely the disappearance of a once-prominent crater and its replacement by the insignificant feature that we find in its place today. This topic will be dealt with in Chapter 6. To the east of Linné the great lunar mountain range known as the Appenines is coming into view. A small crater Conon near the east edge of the Appenines is close to the terminator. It contains details that seem to change position and so has given rise to the suggestion that there are clouds within this formation. North of the Appenines are the Caucasus Mountains and there is a gap separating these two mountain ranges that connects the Mare Serenitatis to the Mare Imbrium on the east, the extreme western part of which is now visible. The crater Cassini, diameter 36 miles, is prominently displayed this evening and is reminiscent of Posidonius. It lies on the Mare Imbrium just south of another great mountain range, the Alps. Cassini is polygonal in outline with broad walls and a narrow crest. The floor is scarcely depressed below the surrounding surface. There is some detail on the floor including the crater A, a smaller crater B and three hills between A and B. Strangely, Cassini was omitted from some of the earlier maps of the Moon. To the south and slightly to the west of Cassini in the Palus Nebularum is the crater Theaetetus, which is 16 miles in diameter. It is notable for its great depth, the floor sinking to almost 5000 feet below the surrounding surface. Almost due west of Cassini and buried in the Caucasus Mountains is the 19-mile crater Callipus. In 1960 a dark spot on the south-east part of its floor was detected by R.A. McIntosh with his 14-inch reflector. I have seen what is presumably the same dark spot with my own eight-inch reflector and it is odd that it is not mentioned by Fauth, Elger, Goodacre or Wilkins and Moore. McIntosh also found a pattern of criss- crossing light streaks on the floor but I have been unable to see these so far.

Fig. 4.13 Cassini. April 6, 1968. Colong. 17.0° app. F.W. Price. Eight-inch refractor. Buffalo Museum of Science, NY, USA

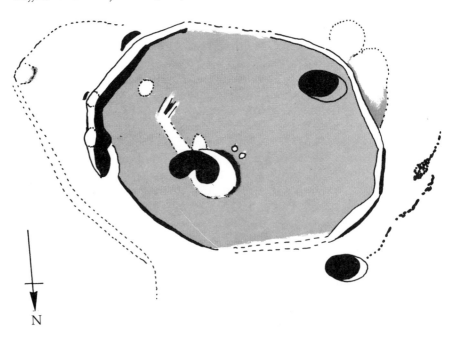

N

Northwards from Cassini is something that looks like a long straight knife cut through the Alps. This is the great Alpine Valley, 80 miles long, and it is one of the finest valleys on the moon.

Moving northwards over the Mare Frigoris we encounter the very large formation W.C. Bond in the light-colored upland terrain. It is a plain bordered by ridges. Its enormous floor is crossed by rows of hills lying generally in a north-west to south-east direction. Between W.C. Bond and the north pole lie the previously mentioned formations Barrow, Scoresby, Main, Challis and Gioja. The north pole itself is situated within a ring designated A, north of the crater Shackleton.

Eighth day

Up to the 'half Moon', or dichotomy, the Moon's terminator is concave but after dichotomy, the terminator becomes convex towards the east and the Moon assumes the shape called 'gibbous', At this eight-day phase more than at any other, the extremely rugged nature of much of the southern hemisphere of the Moon is most vividly evident. The grandest and most striking feature is the great meridionally aligned chain of vast walled plains running down the central meridian of the moon's disc for much of its southern half. The most southerly member of this chain is the walled plain Walter, which has a rhomboidal shape, is 90 – 100 miles in diameter and has very complex ramparts. It is one of the most striking and beautiful objects on the moon under morning illumination when, say, the terminator is about one to two degrees east of the center of its floor. It has a central mountain mass, in the middle of which are two craters.

Moving north we come to Regiomontanus, irregular in shape and also with complex ramparts. It measures 80 miles from east to west and 65 miles from north to south. On the north part of the floor, a little west of center, is a mountain A with a crater on its flank.

To the east of Walter and Regiomontanus is an immense irregular-shaped plain over 100 miles across named Deslandres but formerly known as Hörbiger and sometimes called Hell Plain, presumably because of the 20-mile crater named Hell on its interior. The crater has nothing to do with the nether regions or extreme heat and is named after a Hungarian astronomer. The floor of Deslandres is strewn with craterlets, craterlet chains, craters, hills and many other details too numerous to mention individually but all making for a very interesting spectacle under low angle illumination. The south border of Deslandres is encroached upon by the walled plain Lexell, 39 miles in diameter, the interior of which is decidedly darker than its surroundings. It has a small central hill. Lexell is a fine example of a partial ring, being more or less open on the north so that it looks like a bay on the south border of Deslandres. On the south-east wall of Deslandres is the 25-mile crater Ball, the most remarkable feature of which is a broad deep groove that runs from the crest and down the inner south slope to a crater on its floor lying to the south of the central mountain.

Encroaching on the north wall of Regiomontanus is Purbach, a huge roughly rhomboidal walled plain 75 miles across. Its interior is full of detail

Day 8.

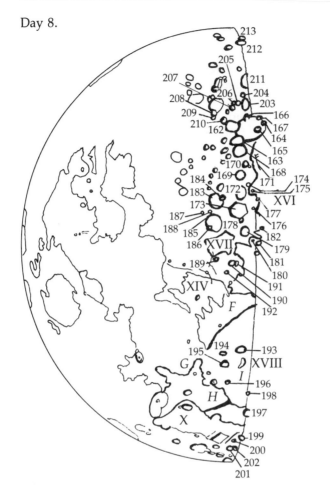

Day 8. Key to formations

162. Walter
163. Regiomontanus
164. Deslandres (Hörbiger)
165. Hell
166. Lexell
167. Ball
168. Purbach
169. Arzachel
170. Thebit
171. Straight Wall
172. Alphonsus
173. Ptolemaus
174. Alpetragius
175. Lassel
176. Palisa
177. Davy
178. Herschel
179. Flammarion
180. Mösting
181. Sömmering

182. Lalande
183. Albategnius
184. Klein
185. Hipparchus
186. Horrocks
187. Halley
188. Hind
189. Triesnecker
190. Murchison
191. Pallas
192. Ukert
193. Archimedes
194. Autolycus
195. Aristillus
196. Mount Piton
197. Plato
198. Mount Pico
199. Birmingham
200. Epigenes
201. Goldschmidt

202. Anaxagoras
203. Orontius
204. Saussure
205. Huggins
206. Nasireddin
207. Miller
208. Furnelius
209. Kaiser
210. Nonius
211. Maginus
212. Moretus
213. Short

   XIV. Mare Vaporum
   XVI. Mare Nubium
  XVII. Sinus Medii
 XVIII. Mare Imbrium

I. Spitzbergen Mountains

including a ridge nearly in the center, which is apparently all that remains of a once complete ring. The north wall of Purbach is interrupted by the crater G. Due north at a distance slightly greater than the diameter of Purbach is the somewhat smaller formation Arzachel, which is 60 miles across. It has massive complex terraced walls and on its floor is a central mountain 4900 feet high. A deep crater designated A is situated on the west part of the floor. There are over 25 hills and craterlets on the floor of Arzachel.

Fig. 4.14 Thebit. November 14, 1964. Colong. 36° app. F.W. Price. Eight-inch refractor. Buffalo Museum of Science, NY, USA

N

Just to the east of the gap separating Arzachel and Purbach is the crater Thebit, 30 miles in diameter. Intruding upon its north-east wall is a much smaller crater which in its turn is intruded upon by an even smaller crater on its north-east wall. This trio of overlapping progressively smaller craters has been amusingly likened to a family of grandfather, father and son. Due east of Thebit on the Mare Nubium is a formation which is virtually unique on the Moon's surface. This is the Straight Wall, an almost straight cliff running in a roughly north-east to south-west direction for about 60 miles, the cliff face looking eastwards. It appears to be a fault, the crust on the east side having sunk considerably below the level of that on the west. At its south end is a curiously shaped group of mountains that has been likened to a stag's horn.

North of Arzachel is the splendid pair of walled formations, Alphonsus and Ptolemaus near the center of the Moon's disc. Alphonsus is 70 miles in diameter and its walls are very broad and complex. It has a central mountain about the height of Vesuvius, which is part of a longitudinal ridge. The floor is full of interesting detail, among which are some dusky spots that are most prominent under a high light. They are said to be 'variable'. Three of these spots are on the west part of the floor and are connected by a winding cleft. A fourth larger triangular spot abuts on the east wall. There are over 50 craterlets on the interior. On the evening of November 3, 1958, the Russian astronomer N.A. Kozyrev was observing the central mountain of Alphonsus using the 50-inch reflector of the Crimean Astrophysical Observatory and witnessed what appeared to be some kind of volcanic activity there; this will be discussed in more detail in Chapter 6.

Just to the east of the gap separating Arzachel and Alphonsus is the 27-mile crater Alpetragius. Its floor is largely occupied by an enormous dome-like central mountain so that under low lighting the entire formation looks rather like a bird's nest with an egg.

East of Alpetragius in the Mare Nubium is Lassel, a low-walled ring 14 miles across with a somewhat irregular outline. It has a small central mountain which is difficult to see. Roughly 20 miles north-east of Lassel is a remarkable group of mountains in association with a bright crater. Continuing in the same direction there is a light oval patch about 10 miles across with a crater on the south edge. This area was described by Mädler as a bright crater five miles in diameter but the present day appearance is definitely not in agreement with this. Ptolemaus is the northernmost member of the great central meridional chain of formations and is one of the most perfect examples of a walled plain on the Moon's surface. In form it is an almost perfect hexagon over 90 miles in diameter and is bounded on all sides by complex discontinuous walls cut through by valleys into separate mountain masses. So large is Ptolemaus that if we stood on the middle of the floor, we would not be aware of being inside a walled plain because the walls would be out of sight below our horizon, except towards the west where one of the wall peaks would be visible. The floor has a dark grey color at sunrise but the tint lightens as the illumination angle increases. The floor looks smooth at a glance but under very low angle illumination it is seen to be covered with a multitude of shallow saucer-like depressions, especially on the east side. The most prominent object on the vast interior is the 4½ mile crater Lyot and there are a few smaller ones. To the east of Alphonsus and Ptolemaus are the two ring formations Palisa and Davy, in a roughly north–south alignment. The more northerly, Palisa is an obscure ring with quite low ramparts and with some craterlets on its interior. Davy is a deep 20-mile crater with a low central mountain. There is a gap in the north wall and a crater intrudes on the south-west.

Fig. 4.15 The Straight Wall. June 18, 1983. Colong. 10.6° F.W. Price. Eight-inch reflector

N

Just outside the north wall of Ptolemaus is Herschel, 28 miles in diameter, and to the north-east of Herschel is Flammarion, a dark-floored plain enclosed by broken walls. Flammarion stands near the south-west border of the Sinus Medii. The brilliant crater Mösting A stands just outside the east wall of Flammarion and Mösting itself is to the north-east and is about 15 miles in diameter. It is very deep with an inconspicuous central mountain. Mösting's companion Sömmering lies to the north-east. It is a ruined ring

Fig. 4.16 Chart of Ptolemaus (based on photographs) (F.W. Price)

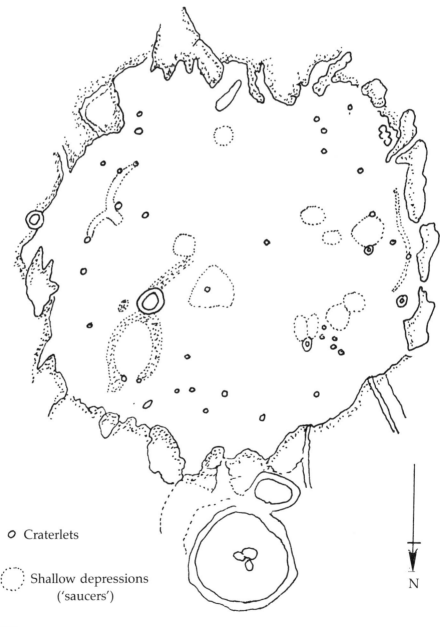

O Craterlets

⊙ Shallow depressions
  ('saucers')

N

17 miles in diameter with low broken walls and a dark interior. Eastwards from Flammarion is Lalande, a very deep crater about 14 miles in diameter. It has bright terraced walls rising to a height of 6000 feet above the floor according to Elger and there is a low central mountain. A long valley outside on the south-west extends as far as the wall of Ptolemaus, a distance of 130 miles.

To the west of the three formations Alphonsus, Ptolemaus and Flammarion are two very large walled plains that are not strictly members of the meridional group. The southernmost is Albategnius, 80 miles across, with massive prominently terraced complex ramparts sliced through with several valleys. The south-east wall is interrupted by the ring formation Klein. The only really prominent feature on the floor of Albategnius is the central mountain, which is over 400 feet high.

Fig. 4.17 Alpetragius. April 20, 1985. Colong. 30.89°. F.W. Price. Eight-inch reflector

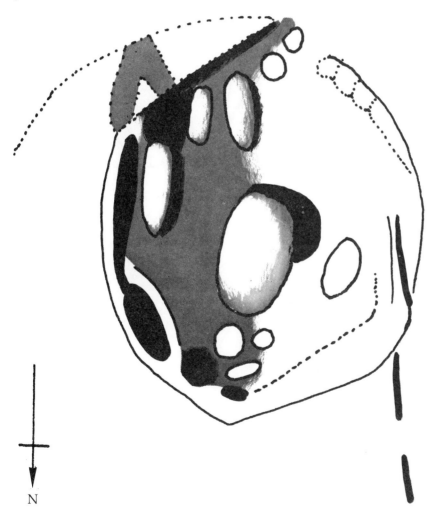

N

To the north is Hipparchus, one of the largest ring formations on the Moon's surface, about 100 miles across. It has very broken and disconnected walls so that it is only prominent under a low Sun. On its floor is the remnant of a low ring. The north-west border of Hipparchus is occupied by the bright crater Horrocks, diameter 18 miles.

Outside the south-west border of Hipparchus is a group of four craters in a curved arrangement. The largest is Halley (21 miles), then Hind (16

Fig. 4.18 Triesnecker clefts (J.N. Krieger)

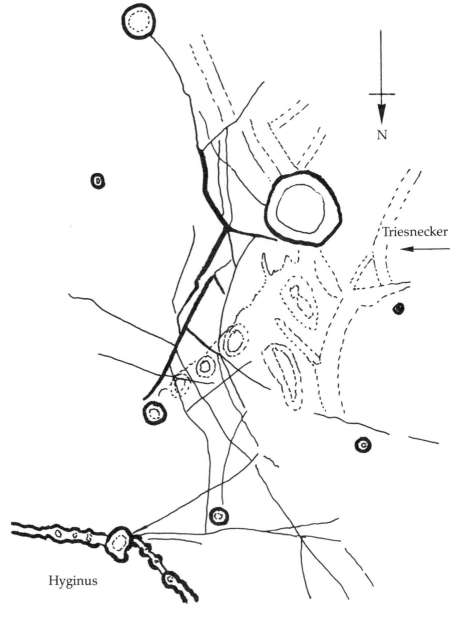

N

Triesnecker

Hyginus

miles), then Hipparchus C and L which are smaller still. This is a good example of what is called a decremental arc, i.e., a curved group of craters of progressively smaller size. There are other examples on the Moon.

The whole of this part of the lunar surface which includes Alphonsus and Ptolemaus is prominently scored by long straight crustal fractures looking like knife cuts, all trending in a north-east to south-west direction.

To the north of Hipparchus in the Sinus Medii is the crater Triesnecker, about 14 miles in diameter and noted for its association with one of the best known systems of clefts on the Moon. Some of these clefts can be seen with a three-inch refractor but the entire complex system requires large apertures and optical excellence to be seen in its entirety. Birt likened these to 'an inverted river system' because several of the clefts seem to become wider and deeper as they reach into higher ground. The entire cleft system lies to the west of Triesnecker; curiously there are none on the east side. East of Triesnecker is the ragged-looking formation Murchison and its companion Pallas. The 35-mile Murchison has complete walls on the west but those on the south look as though they have been broken and are floating away across the Sinus Medii! Pallas, to the east and 30 miles across, is more complete and is a beautiful object in the telescope when the light is right. Its walls are divided by valleys and it has a bright central mountain.

The bright-walled 14 mile crater Ukert lies to the north of Murchison; the most notable feature in this area is the great valley to the east of Ukert over 80 miles long and at least 6 miles wide. It runs in a north-east to south-west direction and rivals the great Alpine Valley.

To the north of Triesnecker lies the Mare Vaporum and further north again we see the Apennine mountains in their entirety. The scarp slope forms a graceful arch that delimits the south-west border of the Mare Imbrium, the largest and most beautiful of the circular maria. The Apennines are the finest range on the visible hemisphere of the Moon. They extend in a practically continuous curve for over 400 miles and include more than 3000 peaks. Everywhere, the Apennines are sliced through by long straight valleys that are part of the general system of crustal fractures that includes those in the Ptolemaus–Hipparchus area.

On the surface of the Mare Imbrium and not far from the terminator on this occasion is the walled plain, Archimedes, the largest of its kind on the Mare Imbrium. The walls are broad and complex and the remarkably smooth interior is about 50 miles across. At sunrise, Archimedes is a splendid sight in the telescope as the west wall casts long spires of shadow across the floor. There are some craterlets on the floor and a system of light streaks. Slightly to the east of the center is a dusky oval area about six miles across.

To the north-west of Archimedes is the fine north-south crater pair, Aristillus and Autolycus. Aristillus, the northern member, is 35 miles in diameter and is the center of a bright ray system. If has beautifully terraced inner walls and the outer walls are scored by deep ravines that are easily visible in small telescopes. On the floor is a fine three-peaked central mountain. An interesting feature of Aristillus is the dark band that runs up the north-west inner slope and extends over the rim. It is actually composed of two streaks that are easier to see outside the crater but a good telescope and fine seeing conditions are needed to see them. They are a good test of telescopic definition and resolution. The walls on the north-east seem strangely bright for about one or two days after full Moon.

Autolycus, 24 miles across, is generally similar to Aristillus. It has wide terraced walls and there is a chain of three craters on the floor. Radiating ridges are seen surrounding it when the illumination angle is low and under a high Sun Aristillus is seen to be the center of a small pale ray system.

On the mare surface to the north of Archimedes are the Spitzbergen Mountains and north-west of these is the isolated mountain Piton.

At this phase of the Moon the Sun is just rising on one of the best known and most observed lunar formations, the walled plain Plato just on the north 'shore' of the Mare Imbrium. It is about 60 miles in diameter and has a smooth grey floor that seems to darken as the sun climbs higher. This makes it a prominent object, lying as it does in the light-colored upland

Fig. 4.19 Archimedes at lunar sunrise. March 8, 1976. Colong. 5.6°. F.W. Price. Eight-inch reflector

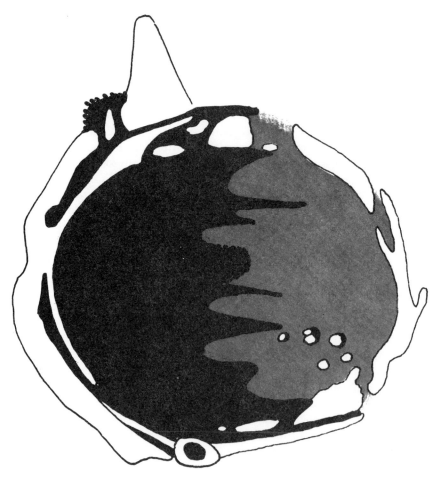

area between the Mare Imbrium and Mare Frigoris. As H.P. Wilkins remarks, 'Everybody who uses a telescope looks at Plato'. There is a tiny craterlet nearly in the center of the floor and two or three others elsewhere that are generally fairly easy to locate under good seeing and lighting conditions. Experienced observers using large telescopes have detected many more and the odd thing is that these craterlets seem to 'come and go' in a curious manner. On occasions when they should be visible they cannot be seen while at other times they are fairly easy objects. The floor of Plato is criss-crossed with delicate whitish streaks which have been carefully mapped but I have never really seen them convincingly, apart from one or two. The south-east part of the crater floor is occupied by a lighter area called the *sector*. Sunrise on Plato is a spectacle not to be missed. At first, when only the crater rim is catching the Sun's rays, Plato looks like a brilliant oval standing out in the blackness. Then, with startling suddenness, a shaft of light appears on the east part of the floor and quickly extends further west. Shortly afterwards others make their appearance. The floor is now crossed with long pointed spires of shadow and some of the most prominent crater-lets are now visible. Gradually the shadows recede uncovering other delicate details. When most of the floor is lit up, the whitish streaks and numerous white spots begin to appear on the dark floor. To the south of Plato on the Mare Imbrium is another isolated mountain, Pico. Across the Mare Frigoris north of Plato are several formations in the light-colored upland area near

Fig. 4.20 Plato at lunar sunrise. January 20, 1975. Colong. 12.1°. F.W. Price. Eight-inch reflector

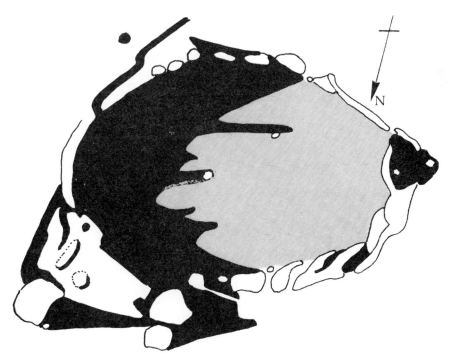

the north pole. Birmingham, which somewhat resembles W.C. Bond, is a very large rhomboidal plain bordered by ridges and containing a nearly central crater B. The whole of this region more than anywhere else on the Moon is characterized by the parallelism displayed by the many large and small formations in this area. To the north-west of Birmingham is Epigenes, 30 miles in diameter, with several craters on its north and south-west crests and many hills on its floor. Between Epigenes and the limb is Goldschmidt, a large ring structure with broken walls and with several craterlets on its light-colored floor. It is intruded upon on the east by the crater Anaxagoras, which is the center of a system of light streaks, some of which cross the interior of Goldschmidt.

We now sweep southwards along the terminator to complete our survey of the eight-day-old Moon. To get our bearings again, we locate Walter and Deslandres. To their south is a group of large formations jostling each other – Orontius, Saussure, Huggins, Nasireddin and Miller.

Orontius is an irregular formation whose once nearly circular shape has been distorted by intrusion of its neighbor Huggins. On its interior are innumerable craterlets and pits; some of these are chains of crater pits that radiate from Tycho. The north part of the floor is occupied by an approximately circular depression that has a smooth floor. Huggins, which encroaches on the west wall of Orontius, is a 42-mile walled plain and its own west wall has been destroyed by intrusion of the crater Nasireddin. The walls of Huggins are not very high and slope gently down to the floor, upon which is what appears to be the remains of a central peak and much other detail. Nasireddin is a fine crater 30 miles in diameter and has regular terraced walls. There is a nearly central crater on the floor. The ring plain

Fig. 4.21 Maginus. Redrawn after K. W. Abineri. Based on observations with eight-inch reflector and photographs. (BAA Moon, 5 (4), 86 (1957))

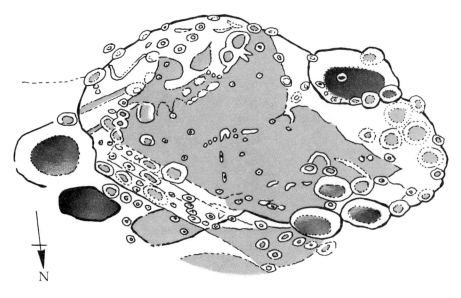

N

Miller encroaches on the north wall of Nasireddin. It is about 30 miles in diameter and has a multiple central mountain. Saussure is a walled depression 30 miles across lying close to the south wall of Orontius. Its bright terraced walls attain a height of 8000 feet on the east. The interior is rather dark and there is a double craterlet slightly west of center. One of the streaks from Tycho is deflected by Saussure and another can be faintly traced on its floor.

To the north-west of this group of formations and lying in the gap between Stöfler and Walter are the three formations Furnelius, Kaiser and Nonius. Furnelius, which makes contact with the north-east wall of Stöfler, is a walled plain 40 miles across. It has a fairly level floor upon which are some small craterlets. On the east Furnelius is overlapped by a similar ring A about half its size. Adjoining Furnelius on the north-west is Kaiser, a large ring on the floor of which are two craters and two chains of craterlets. Nonius is a large formation north-east of Kaiser and impinges on the wall of Walter. It has broken walls and the interior has many ridges and large low rings.

South of Saussure is the splendid formation Maginus, a walled plain 110 miles across. It has ruined walls and its inner slopes are broad. Slightly west of center is a group of hills. There is another mountain group south-west of this and further south again a large ruined ring lies against the inner south wall. Large telescopes reveal multitudes of crater pits, hillocks and craterlets on the floor of Maginus. A group of four large rings make close contact with the outer north and north-west walls and a 30-mile crater breaks into the wall on the south-east. At full moon it is very difficult, almost impossible, to locate Maginus. Indeed, as Mädler said, 'the full moon knows no Maginus'.

Immediately to the south of Maginus is a rugged upland area devoid of sizeable formations and south of this is Moretus, a fine ring structure 75 miles in diameter and seen foreshortened into a flat ellipse owing to its closeness to the limb. The floor is somewhat dark and on it is a splendid central mountain rising to 700 feet. According to Mädler this is the highest central mountain on the Moon. The last formation we will note on this survey of the eight-day-old Moon is Short, a 35-mile ring immediately to the south of Moretus. On its floor is a small bright craterlet with a little hill on its south edge.

Ninth day

This evening we are treated to one of the most striking and beautiful of telescopic objects, the great walled depression Clavius. It is situated in the light-colored upland terrain in the Moon's southern hemisphere and is fully 145 miles across. So large is it that when it is on the terminator it forms a bulge that can be distinctly glimpsed even with the naked eye as an irregularity on the terminator. Its somewhat irregular wall is intruded upon on the south by a 25-mile crater, Rutherford, which has a peak on its interior. From the outer north wall of Rutherford radiate several ridges visible soon after sunrise, three of which are especially prominent. T.G. Elger likens these to the ribbed flanks of some of the volcanoes in Java. The north rampart of Clavius has a similar ring of almost the same size as Rutherford and named Porter. It has a triple central mountain. On the light-colored floor of Clavius is a curving group of four craters, the largest on the west and the others progressively smaller going eastwards, the curve being convex to the north.The second largest of these craters is associated with a

Day 9.

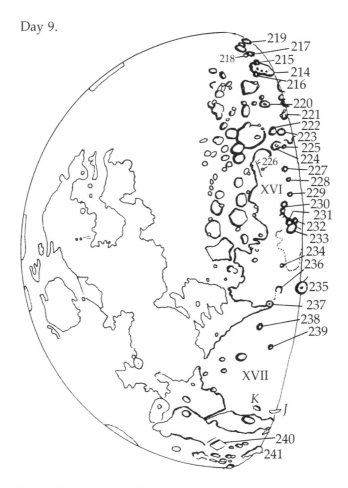

Day 9. Key to formations

214. Clavius	226. Birt	238. Wallace
215. Rutherfurd	227. Wolf	239. Timocharis
216. Porter	228. Gould	240. Bartlett (Mädler's Square)
217. Gruemberger	229. Opelt	241. J.J. Cassini
218. Cysatus	230. Guericke	
219. Newton	231. Parry	XVI. Mare Nubium
220. Tycho	232. Bonpland	XVII. Sinus Medii
221. Heinsius	233. Fra Mauro	
222. Gauricus	234. Gambart	J. Straight Range
223. Wurzelbauer	235. Copernicus	K. Teneriffe Mountains
224. Pitatus	236. Stadius	
225. Hesiodus	237. Eratosthenes	

complex group of hills. This is probably the best example of a decremental crater arc on the Moon's visible surface. Scattered over the floor of Clavius are many craterlets, which are especially numerous in the southern half of the floor. Oddly, no rills or clefts have ever been detected on its vast interior. One never tires of studying Clavius, there always seems to be something that we missed seeing on a previous occasion. It appears to have an inexhaustible supply of detail. Sunrise on Clavius is a most spectacular sight, especially when the Sun's light is just catching the rims of the interior craters and they and the rim of Clavius itself stand out brilliantly against the darkness.

Between Clavius and the south pole there are not too many formations of note. Gruemberger and Cysatus are situated closely against the north wall of Moretus. These two are members of the group of formations including Newton and Short, of which Moretus is the chief. The 58-mile Gruemberger has its outer north-west wall in common with Cysatus and is roughly pear-shaped. Cysatus, 28 miles in diameter, has a low central mountain on its rather small floor which is somewhat difficult to detect.

Due north of Clavius is the crater Tycho, best known as being the apparent center of the most prominent and extensive system of bright rays on the Moon. The rays are most prominent under a high Sun and for this reason Tycho is the most prominent of all lunar formations at full Moon and has earned the title of the 'Metropolitan Crater of the Moon'. The rays must be some sort of thin surface deposit as they cast no shadow. They stretch over the lunar surface for hundreds of miles running across valleys and mountains without apparent hindrance or suffering any deviation from their course. Tycho is 56 miles in diameter and has a prominent central mountain on its floor and a crater to the west of it. T.G. Elger remarks that for some inexplicable reason the floor of Tycho never looks very distinct. The walls of Tycho are composed of many linear segments and the entire formation is roughly circular. On the inner north-east slope Gaudibert delineated a cleft-like object. The country immediately outside Tycho is somewhat darker than the surrounding light-colored terrain and forms a 'collar' around the crater. Everywhere is peppered with multitudes of pits arranged in radiating chains.

North-east from Tycho and close to the terminator is Heinsius, a curious formation 45 miles across, chiefly notable for the two large rings B and C intruding on its south-east and the ring A on its floor. The three similar-sized rings make a neat equilateral triangle.

To the north-west of Heinsius is a fine pair of large walled formations aligned in an east–west direction. These are Gauricus (40 miles) and Wurzelbauer (50 miles). Gauricus has an irregular border and a large ring adjoins it on the south. Apart from crater pits and ridges there is no prominent detail on the floor. Wurzelbauer is nearly circular and has a very complex border. On its uneven floor there is much fine detail, mostly craterlet chains and rows of hills.

Close to Gauricus and Wurzelbauer is the magnificent ring plain Pitatus, situated on the south 'shore' of the Mare Nubium. It is 50 miles in diameter and looks very much as though it has suffered from erosion in the past.

Its northern 'seaward' wall has wide gaps. The dark-colored floor looks as though it was formed by flooding with molten lava from the Mare Nubium at some time in the remote past. There is a not quite central hill and two whitish patches on the southern part of the floor and another on the north. A cleft can be followed all around the interior close under the walls. There are other clefts on the north-west and north-east parts of the floor. On the south-west of Pitatus there are two close parallel rows of crater-like depressions, possibly the most noteworthy of their kind, that extend as far as the west flank of Gauricus.

Closely associated with Pitatus is the smaller formation (28 miles) Hesiodus close by its outer east wall. It communicates with Pitatus via a pass. The ring A close against the outer north-east wall contains a bright inner ring perfectly concentric with the walls of A.

Travelling northwards across the Mare Nubium we pass the crater Birt some distance to the west near to the Straight Wall. It was hidden in the darkness beyond the terminator on the previous evening. Its wall is broken by a smaller crater A on the west. There are two dusky bands running up the east wall of Birt and on the mare surface to its east is a fine curved cleft.

Fig. 4.22 Chart of Clavius. From a photograph and personal observations with eight-inch reflector. F.W. Price

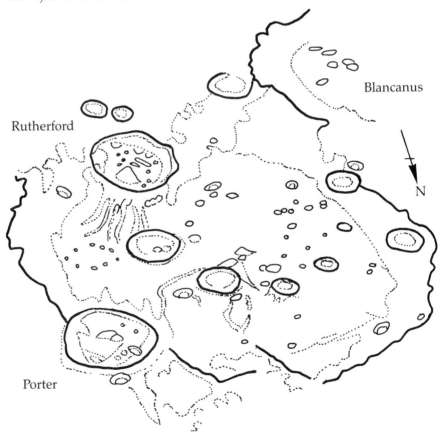

North from Pitatus are three somewhat fragmentary formations in merid-ional arrangement, Wolf, Gould and Apelt, and south of these is the interest-ing group of four formations, Guericke, Parry, Bonpland and Fra Mauro. Guericke stands apart from the other three which press closely against each other. Guericke, 36 miles across, is a partially destroyed ring plain. Its walls are especially broken on the west and north and the south-west wall is a curving group of separate blocks. There is much detail within Guericke, the most prominent of which is a crater lying close under the east wall. East of the incomplete rings to the north of Guericke extends a line of tall mountain masses to the west side of Parry and for 30 miles further north. These reminded T.G. Elger of the avenues of monoliths constructed by Druids.

Parry is a 28-mile ring plain with fragmentary walls but more complete than Guericke. There is a nearly central crater on the floor and a prominent crater on the east wall. South of Parry is the crater A from which originates a cleft that travels across the floor of Parry close to the east wall and then passes into Fra Mauro and crosses over the floor of this formation.

Fra Mauro is the largest of this group of three formations and has part of its south-east wall in common with Parry. It is a ruined formation roughly 50 miles across. The floor is variegated in hue and there are many small craters and ridges. The entire north-west portion of the interior is occupied by a dark patch which has finger-like projections pointing southwards. A long cleft traverses the floor north and south of the central crater A and

Fig. 4.23 Pitatus. Redrawn after W. Goodacre F.W. Price

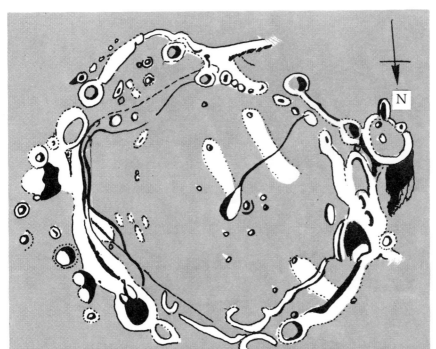

there are others. Bonpland to the south of Fra Mauro has its north wall in common with the latter and its west wall in common with Parry. It also has a ruinous appearance but is a fine example of a ring plain in spite of this. There are some craterlets and hillocks on the interior, which is crossed by three clefts. On the north of a discontinuous ridge running in a northerly direction from Fra Mauro is the rather odd-shaped ring of Gambart, 16 miles in diameter and situated nearly on the lunar equator. It has a level floor with a light patch on the south and a couple of low swellings in the north. There does not appear to be a central peak.

Moving further north we are confronted with the magnificent formation Copernicus, the finest example of a crater on the Moon's surface, standing in grand isolation in the Oceanus Procellarum. T.G. Elger aptly dubbed it 'the Monarch of the Moon'. It is not too far away from the apparent center of the Moon's disc and so is seen undistorted by extreme foreshortening, unlike other formations such as Petavius, which would be fully as splendid

Fig. 4.24 Hesiodus A showing the concentric inner ring. December 23, 1966. Colong. 39.7°. From a photograph. F.W. Price

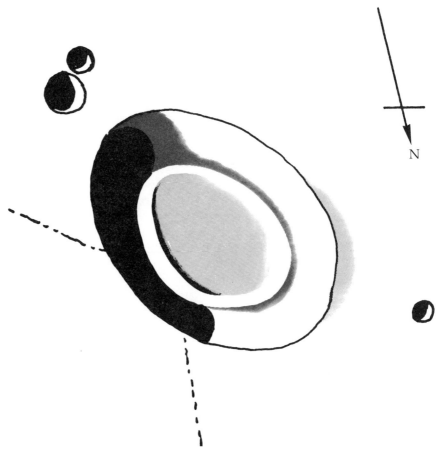

as Copernicus if they were not so close to the limb. Copernicus is 56 miles
in diameter and although approximately circular its massive complex walls
are made up of several linear segments so that the outline is decidedly
polygonal. The walls rise to a sharp thin crest 12 000 feet above the floor
and their gentle outer slopes consist of many concentric ridges. Many radial
ridges spread out from the ring and among them are scattered multitudes
of craterlets and small pits. The nearly circular floor is 40 miles in diameter
and almost in the center are five bright mountains, according to Elger.
Wilkins and Moore mention seven peaks. When viewed with high power
and good steady seeing, sunrise on Copernicus is a sight of unearthly
splendor.

Another notable feature of Copernicus is that it is the apparent center
of the second most prominent and extensive ray system on the Moon. The
rays are not as bright as those of Tycho and instead of being well-defined
and straight are wispy and plume-like. Some of the rays when traced back
towards Copernicus are tangential to the wall and in at least one case
completely by-passes the crater.

West of Copernicus is the curious low ring of Stadius, which looks as
though it is old and much eroded. The walls are very fragmentary, the
north wall being a mere 'ghost' marked by crater pits. The 40-mile interior
is dotted all over with crater pits, and delineations by various observers
differ from one another. Impinging on the north-west rim of Stadius is the
southern tip of a mountain arm extending southwards from Eratosthenes,
a large crater situated at the extreme southern tip of the Apennine mountains. It may be fairly described as a smaller version of Copernicus. It is 38
miles in diameter and is a striking telescopic object when the illumination
angle is low. It is difficult to locate under a high Sun and all that can be
seen is a confusion of light and dark patches. At sunrise when the floor is
about half filled with shadow, the many irregular terraces on the inner east
wall are well shown. There is a complex central mountain which has a
crater-like depression on it. As the lunation progresses, dusky streaks and
patches appear within Eratosthenes which seem to shift position from night
to night in a curious and inexplicable manner.

About half of the Mare Imbrium is exposed at this phase and scattered
on its surface are several interesting medium-sized craters standing in isolation. Due north of the extreme southern tip of the Apennines is a low
ring named Wallace. It seems to have been partially filled in by mare material
ages ago when the surface materials were still fluid. There is little to be
seen on the floor except some tiny specks, one of which is plainly a craterlet
and is situated nearly centrally. The extreme southern part of the wall is
an odd-looking triangular projection which descends to the mare level, the
tip being hooked towards the north.

Timocharis, a perfect crater 25 miles across, is some distance due north
of Eratosthenes and a little south of east from Archimedes. It has broad
walls that are finely terraced on the inside and on the floor is a central
crater, a few mounds and low hills. Timocharis looks very bright and 'fresh'
under a high light and is the center of an extensive but faint ray system.
There are other interesting craters on the Mare Imbrium but they are hidden

in the darkness beyond the terminator and we will have to wait until tomorrow night to see them.

We have to sweep a considerable distance northwards over the Mare Imbrium almost to its north 'shore' before encountering anything noteworthy. Here, just peeping out of the blackness is the range of mountains

Fig. 4.25 Birt and cleft (Mount Wilson photograph). Redrawn after E. A. Whitaker (BAA Moon, 4 (4), 75 (1956)

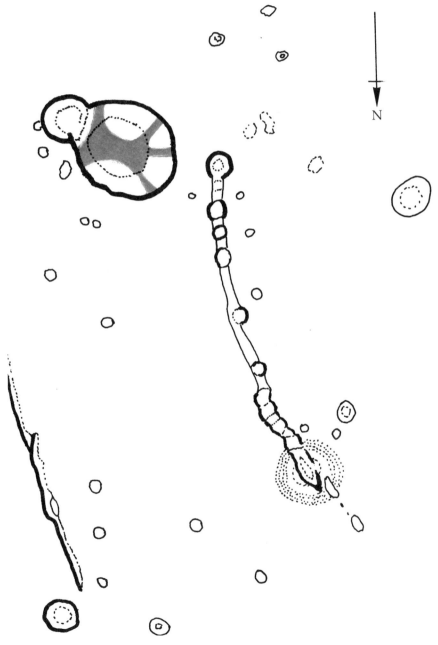

called the Straight Range and to their west are the Teneriffe Mountains close to Plato. The next interesting object lies across the Mare Frigoris on its north 'shore'. Here, immediately to the west of the crater Fontenelle is an ill-defined roughly diamond-shaped area that Mädler drew with startling prominence and geometrical regularity on his map. This area, nowadays

Fig. 4.26 Sunrise on Copernicus. After Nasmyth

N

known as Bartlett, was once called Mädler's Square. This is another part of the Moon where physical change is supposed to have occurred but we will defer discussion of this until Chapter 6. The north-west boundary of Bartlett is bounded by the wall of Birmingham.

North of Fontenelle is the large plain bounded by ridges, called J.J. Cassini and named after the son of the other astronomer of that name, also immortalized by having a crater in the Mare Imbrium named after him.

Tenth day

The observer's eye is immediately attracted by a beautiful silvery sickle of light extending out from near the 'bottom' (north) part of the terminator in the blackness beyond. This is the curved wall of the Sinus Iridum. The Sinus appears to be the remains of a once complete ring, of which the south wall has now completely disappeared, probably inundated by the fluid lava of the Mare Imbrium in the remote past. It is approximately semicircular and its curved wall, now called the Jura Mountains, has a cape at each end. The eastern cape is named Heraclides and under certain lighting conditions it bears a striking resemblance to the profile of a woman's head with long flowing hair. The apparition is often called the Moon Maiden. She looks across the bay to the other cape named Laplace, which also sometimes has the appearance of a feminine facial profile. Cape Laplace casts a prominent shadow that persists for several days after sunrise. On the floor of the bay are some ridges, craterlets and hillocks. Under a high light some ghostly rings become visible, these probably being the sites of sunken ring formations.

Scattered over the Mare Imbrium are several isolated moderate-sized craters. To the south of Cape Laplace are the Leverrier–Helicon pair, respectively 11 and 13 miles in diameter. Helicon has a central crater and Leverrier a central hill. Leverrier is a dark crater under low illumination and is sometimes difficult to find at full Moon.

Some distance due south is Lambert, 18 miles in diameter. It has broad walls finely terraced on the inside and a central craterlet standing on a long ridge. At sunrise a nearly continuous valley may be seen running round the outer west slope. Moving north again we come to the smaller formation Pytheas, a rhomboidal ring plain 12 miles in diameter. It is very bright, has a brilliant central peak and is the center of a light ray system.

Near the terminator roughly south-east of Lambert is the 19-mile crater Euler also the center of a minor ray system. Moving south along the terminator we come to Tobias Mayer, a beautiful 22-mile crater situated at the east end of the Carpathian Mountains, which lie to the north and north-east of Copernicus. The Carpathians and Tobias Mayer are a beautiful telescopic spectacle at sunrise. At the west end of the Carpathians is the crater Gay-Lussac, due south of Copernicus.

Travelling back 'down' (northwards) the terminator, we see the little eight-mile crater Caroline Herschel, named after the sister of the famous English astronomer Sir William Herschel, just peeping out of the darkness.

Plate 1. *The Moon at last quarter*

Plate 2. *Posidonius and the Serpentine Ridge. Evening illumination*

Plate 3. *Mare Nectaris, Altai scarp, Fracastorius, Piccolomini, Theophilus, Cyrillus, Catharina. Sunset*

Plate 4. *The surface from Stofler—Maurolycus to the Altai scarp. Sunset*

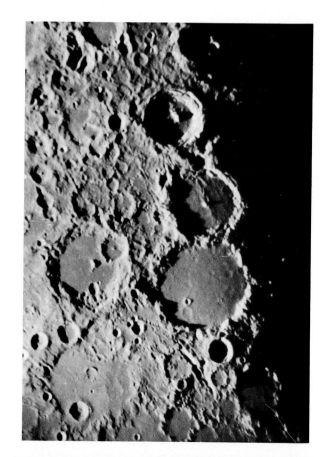

Plate 5. *Arzachel, Alphonsus, Ptolemaus, Albategnius, Hipparchus. Sunrise*

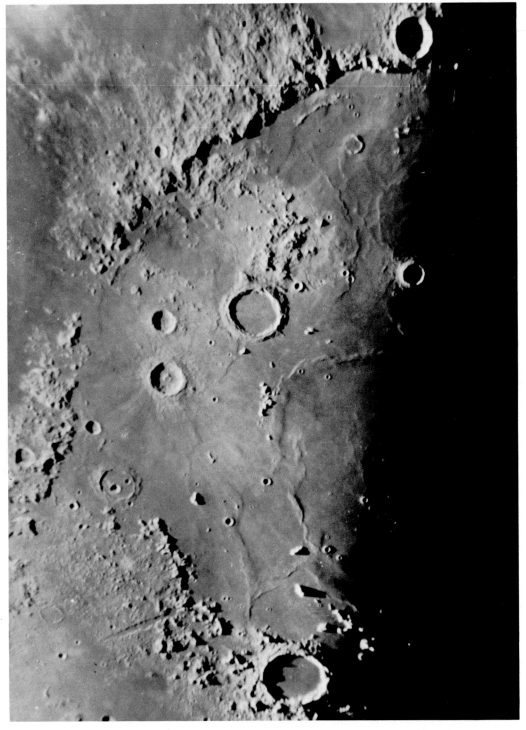

Plate 6. *The Lunar Alps, Apennines, Caucasus Mountains and Mare Imbrium. Sunrise*

Plate 7. *The south polar region at lunar sunrise showing Clavius and Maginus*

Plate 11. *Sinus Iridum and John Herschel. Sunrise*

Plate 8. *Clavius,
Maginus and
Tycho. Sunset*

Plate 9. *Copernicus,
Eratosthenes and
Stadius. Sunrise*

Plate 10.
*Copernicus and the
Carpathian
Mountains. Sunrise*

Plate 12. *Mare
Humorum,
Gassendi, Letronne
and Doppelmayer.
Sunrise*

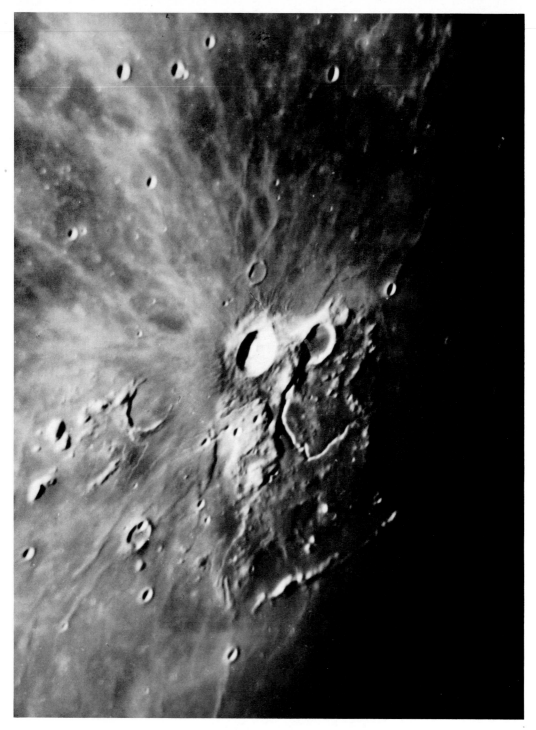

Plate 13.   *Aristarchus, Herodotus and Schröter's Valley, with Prinz and the Harbinger Mountains. Sunrise*

To the north of the Cape Laplace in the light-colored uplands are two notable formations, Maupertius and Condamine. Maupertius is irregular in shape and about 20 miles across and is not easy to describe because of its complex nature. Condamine, 30 miles in diameter, is situated on the south border of the Mare Frigoris. It has a rhomboidal shape. Extending west from Condamine and all the way to Plato is a curious and interesting

Day 10.

Day 10. Key to formations

242. Cape Heraclides
243. Cape Laplace
244. Le Verrier
245. Helicon
246. Lambert
247. Pytheas
248. Euler
249. Tobias Mayer
250. Gay Lussac
251. Caroline Herschel
252. Maupertius
253. Condamine
254. Philolaus

255. Reinhold
256. Lansberg
257. Bullialdus
258. Lubiniezky
259. Kies
260. Mercator
261. Campanus
262. Capuanus
263. Cichus
264. Longomontanus
265. Wilhelm I
266. Montanari
267. Lagalla

268. Blancanus
269. Scheiner
270. Klaproth
271. Casatus

XIX. Sinus Iridum
XX. Palus Epidemiarum

L. Jura Mountains
M. Carpathian Mountains
N. Riphaen Mountains

region which consists of a plateau upon which are several small hills and many craters and craterlets, many of which are arranged in chains.

Between here and the limb, beyond the north shore of Mare Frigoris the only formation worthy of note is Philolaus, a 46-mile crater to the north-east of Fontenelle with terraced walls that attain a height of 12 000 feet. It overlaps an old ring to the east.

After this somewhat erratic tour of the Mare Imbrium and the region to the north, we now swing the telescope southwards again to the dark marial region south-east of Copernicus where we come across two fine isolated craters, Reinhold and Lansberg. Reinhold, 30 miles in diameter, has walls rising on the west to 9000 feet above the floor. There is a very distinct regular terrace on the inner slope of its wall, that on the east side being well seen when the interior is about half lit at sunrise. From the south-west wall runs a fine mountain ridge and there is another much shorter arm extending from the north wall. These and other details in the area are an interesting sight in the telescope. Lansberg, which is similar to Reinhold and just a little smaller, has an enormous crescent-shaped terrace on its inner east slope, which appears to be a land slip. It is separated from the wall by a ravine, which is filled with shadow soon after lunar noon. There is a central mountain mass with multiple peaks. T.G. Elger mentions that he was observing sunrise on Lansberg on January 23, 1888, using his 8½-inch Calver reflector when the floor was about three quarters covered with shadow. He noticed that the illuminated part was of a dark chocolate color that contrasted strongly with the grey color of the surrounding area. This

Fig. 4.27 'Ghost' rings in Sinus Iridum. Redrawn after W. Goodacre

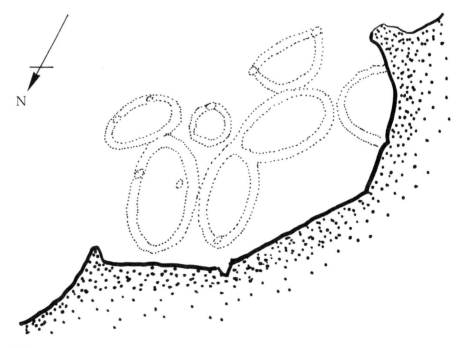

N

Fig. 4.28 Interior of Bullialdus. March 5, 1982. Colong. 38.8°. F.W. Price. Eight-inch reflector

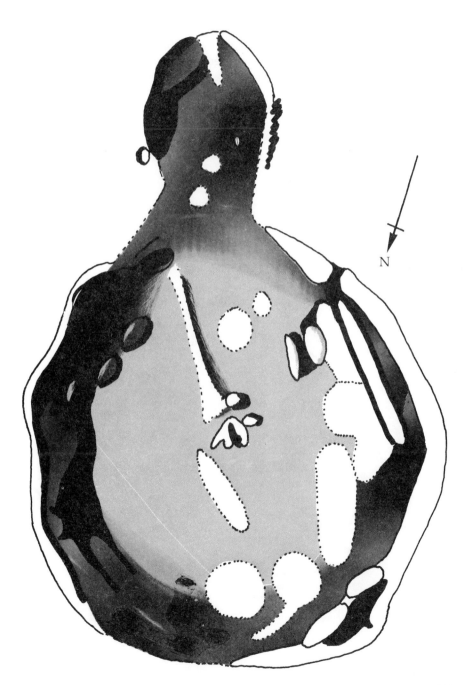

N

persisted until over half the floor was sunlit and thereafter gradually became less pronounced.

To the south of Lansberg are the Riphaen Mountains and the Urals. The Riphaens are very bright and extend for about 100 miles in a north – south direction. They seem to be the remains of what were once prominent ring plains that have been fragmented.

Some distance south of the Riphaens is one of the finest of the lunar ring plains, Bullialdus, 39 miles in diameter and the most notable object in the Mare Nubium. It is a very 'typical' lunar crater, and in many ways is a miniature of Copernicus. Like Copernicus, there are ridges radiating in all directions on the outside. There is a fine compound central mountain of four peaks and a floor which is concave according to Elger but which Wilkins and Moore think is convex. What appears to be an ill-defined ridge runs from the central mountain across the floor to the south wall. The inner slopes are magnificently terraced. On the inner east slopes are two deep parallel terrace valleys which are best seen when the morning terminator

Fig. 4.29 Kies. April 10, 1984. Colong. 30.7° . F.W. Price. Eight-inch reflector

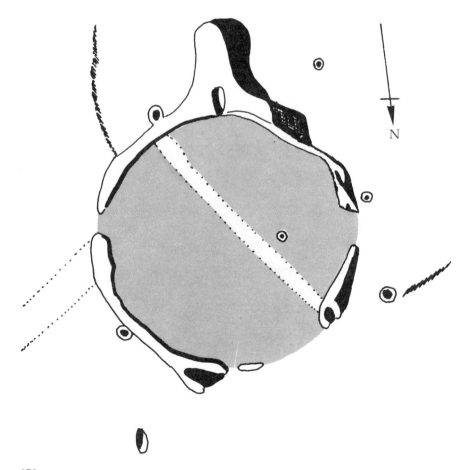

Fig. 4.30 Domes in Capuanus. May 13, 1981. Colong. 30.3°. F.W. Price. Eight-inch reflector

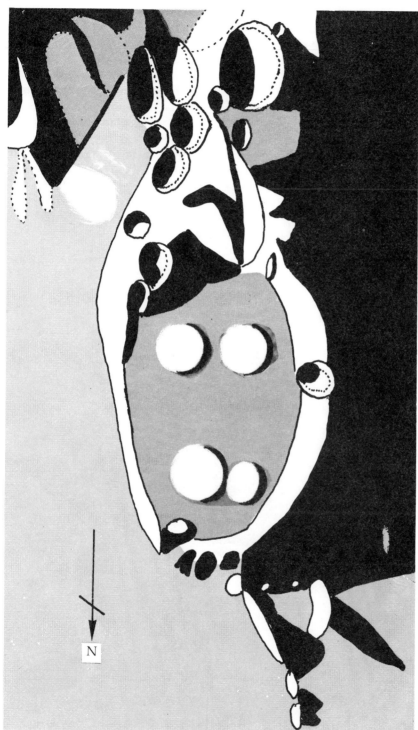

lies on the west border of the Mare Humorum. On the inner east and south-east slopes are several bright objects that look like low dome-like structures under morning illumination. They are quite distinct but none of the maps or drawings I have yet seen of Bullialdus depicts them really accurately so I have observed them and made several careful drawings of these objects. The floor of Bullialdus is dull under low light and is evenly shaded but under high illumination it is bright and several dark areas are to be seen. One of these, north-east of the central mountain, seems to enlarge in a clockwise direction in the lunar morning and spawns dark areas in the west and north-west parts of the floor, according to Wilkins and Moore. These all disappear as sunset draws near. Similar appearances have been noted in other formations. It is probable that the floor of Bullialdus is quite rough so that at sunset all the floor is covered with tiny shadows. The tiny features casting these shadows are not resolvable in the telescope and their presence is indicated only under low angle illumination when their shadows combine. Just outside Bullialdus to the south-west is the crater A, connected to Bullialdus by a shallow valley. Another crater B

Fig. 4.31 Klaproth and Casatus. March 2, 1966. Colong. 40.0°. F.W. Price. Eight-inch refractor. Buffalo Museum of Science, NY, USA

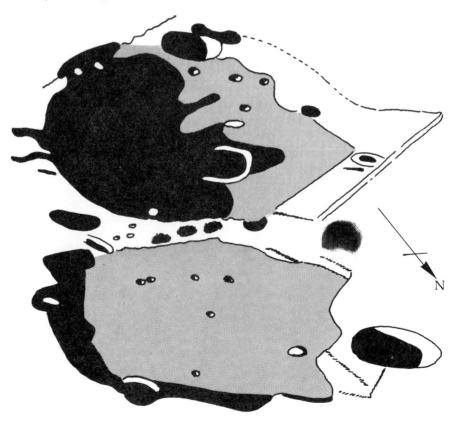

further away and to the north of Bullialdus is likewise connected to it by a wide shallow valley. To the south of Bullialdus is the low ring of Kies, 25 miles in diameter. It has a mountain spur projecting from the outside of its south wall and its floor is traversed by a light streak extending from Bullialdus in a south-west direction. To the east of Kies is a single prominent dome. South of Bullialdus is the ruined ring of Lubiniezky, 24 miles in diameter. Its walls are broken on the south and a bright streak crosses the floor. One of the rays from Tycho passes a little to the south-east of Lubiniezky.

To the south-east of Kies is the close crater pair Mercator and Campanus, both about 29 miles in diameter. On the rather dark floor of Mercator are some craterlets. The south wall is sliced by a cleft which curves over the floor and ends in a depression near the center and just inside the crest of the east wall are two craterlets. Campanus has a darker floor than Mercator and on it are a central hill, two incomplete rings with their concave faces to the west, a crater chain on the north and a couple of craterlets between the north wall and the central mountain. A mountain arm projects from the north wall.

To the south of Campanus on the south 'shore' of the Palus Epidemiarum is Capuanus, 35 miles in diameter, whose floor is higher than the surrounding country. It is most notable for the large low domes on its dark floor. They are easy objects in an eight-inch reflector at sunrise. The floor is crossed by light streaks running in a north–south direction. The continuity of the south wall is destroyed by many large circular formations and deep valleys. From its east and north-east walls, three arm-like projections stretch out over the Palus Epidemiarum.

West of Capuanus is Cichus, 20 miles in diameter, situated on the west 'shore' of the Palus Epidemiarum. There is a prominent deep crater about five miles across designated G on its east rim, which was depicted larger by Mädler than by Schröter, which caused Webb to think that a physical change may have occurred there. On the floor of Cichus to the west of center are two low hills. One of the bright rays from Tycho runs close to the west wall of Cichus.

Some way to the south of Cichus and north of Clavius is the great 90-mile walled plain Longomontanus, one of the largest on the Moon. Its walls are much broken by depressions, mostly on the north-east, and on its dusky floor near the center is a group of hills, three of which are larger than the rest. The floor is crossed by rays from Tycho. Immediately to the north are three other closely grouped formations, Wilhelm I, Montanari and Lagalla. Wilhelm I is an irregular ring 60 miles across and has a rough uneven floor. It is crossed by three bright rays from Tycho in a west to east direction, the two on the north being especially bright under a high Sun. Montanari has a craterlet on its floor and an odd-looking pointed enclosure on the west. Lagalla is pear shaped and has broad terraced walls. Close to the south-east and east wall of Clavius are the two large walled plains Blancanus (57 miles) and Scheiner (70 miles). Blancanus has broad walls of no great height and on its interior is a central group of three peaks. There are three prominent craters between these and the inner south slope, hills and craterlets. The floor itself is light grey and the inner walls are bright. Scheiner has prominently terraced walls and there are seven craters on its floor.

The last two formations to claim our attention this evening lie south of Blancanus. These are Klaproth and Casatus. Klaproth is a walled plain 60 miles in diameter. On its interior are some low ridges and several craterlets, difficult to detect, although I have seen some of them with an eight-inch refractor. Casatus, 70 miles across, intrudes slightly on the south-east wall of Klaproth. It is a fine ring plain with walls rising to 18 000 feet on the west and a peak rising to 22 000 feet above the floor on the east. At the bottom of the inner south-east wall is a row of elevations that look like huge boulders and there is a deep crater C on the north part of the floor. On the same evening that I saw the craterlets on the floor of Klaproth, this crater C was hidden by the shadow of the west wall of Casatus except for part of its rim which appeared as a bright horseshoe of light standing out from the shadow.

Eleventh day

The continuing eastward movement of the terminator has now brought into view the last of the marial bodies that are successively displayed during the first half of the lunation; this is the Mare Humorum, nearly but not quite isolated from the vast expanse of the Oceanus Procellarum. In some ways it is almost the eastern counterpart of the Mare Crisum. The Mare Humorum is nearly circular but is seen foreshortened into an ellipse owing to its comparative nearness to the limb. The dimensions are given by E. Neison as 263 miles north to south and 286 miles east to west. It has well-defined borders and a dark floor, which makes it a prominent feature at full Moon, even to the naked eye. On the surface of the Mare are numerous craters, craterlets and crater pits, the largest of which is about eight miles in diameter. Most of these craterlets are seen as white spots under a high light. Under a low Sun, a number of ridges concentric with the Mare border can be seen on the east and west sides of the surface.

On the north 'shore' is one of the most beautiful and intensively observed lunar formations, the walled plain Gassendi, about 55 miles in diameter. It has a curious lop-sided appearance and its walls are quite high and complete except on the extreme south where there is a gap. It seems that in the remote past the fluid material of the Mare Humorum destroyed this part of the south wall of Gassendi and flooded the south part of the floor. At the center is a splendid group of mountain peaks, the highest of which rises to 4000 feet. On the south-east part of the floor is the remnant of what was probably once a nearly concentric interior ring. Gassendi is notable because of the variety of details on its interior, among which are about 30 or 40 clefts. Two of the most prominent of these diverge from the central mountains over the floor in a south-west direction. Associated with these are two craterlets that are very bright at full moon. These and other clefts in the south-west part of the floor are easily visible in a four-inch telescope but the others that are on the east half of the interior need large apertures to be seen. They criss-cross or run parallel in a complex manner and cannot all be seen on any one occasion. Good seeing, the right angle of illumination and favorable libration are essential for their successful visualization. The cleft system has been charted by different observers and there is a strange diversity among the different delineations. In fact, it has been said: 'Every Man his own Gassendi'. The north wall of Gassendi is intruded upon by the crater Clarkson and just to the east the wall of Gassendi is sliced through

Day 11.

Day 11. Key to formations

272. Gassendi	283. Weigel	294. J. Herschel
273. Clarkson	284. Letronne	295. Anaximander
274. Hippalus	285. Flamsteed	296. Carpenter
275. Vitello	286. Kepler	297. Anaximenes
276. Agatharchides	287. Encke	
277. Doppelmayer	288. Diophantus	XXI. Mare Humorum
278. Lee	289. Deslisle	XXII. Oceanus Procellarum
279. Hainzel	290. Bianchini	
280. Mee	291. Sharp	P. Percy Mountains
281. Schiller	292. Mairan	
282. Röst	293. Harpalus	

by two prominent gaps that continue across the floor as two parallel clefts. On the east wall is a curious crater-like depression from which extends on the outside in a south-east direction the Percy Mountains defining the north-east border of the Mare Humorum.

The western 'shore' of the Mare Humorum is marked by a group of three prominent concentric clefts associated with the 38-miles Hippalus and they are prominent enough to be seen in a four-inch telescope. Their origin is

Fig. 4.32 Clefts in Gassendi. (Based on Orbiter V photograph) F.W. Price

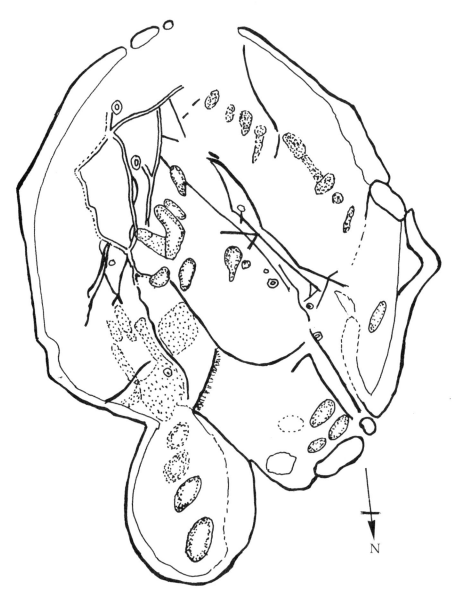

N

in the rough country east of Campanus according to Elger but Wilkins and Moore state that they diverge from a point further south near the crater Vitello on the south 'shore' of the Mare Humorum. The most easterly cleft cuts right through Hippalus. Hippalus itself is another example of an incomplete ring, the wall of which has been destroyed on the side facing the Mare Humorum. North of Hippalus is Agatharchides, an irregular formation 30 miles across. It is a complex structure and needs to be observed under different lighting conditions before a clear idea of its distinctive features can be obtained. The south wall is cut through by parallel valleys trending towards Hippalus and on the floor is a low central mound. Across the Mare Humorum opposite to Agatharchides is the 40-mile ring of Doppelmayer, a fine example of its kind. The part of the ring wall lying on the Mare surface has almost entirely disappeared. Under low lighting a broad low ridge can be seen extending across the opening in the wall. At the south-west end of the ridge is an isolated mountain that casts a long pointed shadow onto the floor at sunrise. A prominent central mountain 2500 feet high stands in the center of the floor and is cut in two by a ravine.

On the south 'shore' of the Mare Humorum are the smaller craters Vitello and Lee. Vitello, a fine crater 30 miles across, has as its most interesting feature a bright inner ring not quite concentric with the main wall. The boss-like central mountain within this ring has been likened to the Matterhorn and at lunar sunrise the inner ring has been reported by F.H. Thornton to look like an incomplete chain of beads around the central mountain. To the east of Vitello is the similar-sized incomplete ring structure Lee. On its floor are some details including a ring on the west which, though easy to see, was strangely not recorded by W. Goodacre. There is a cleft discovered by Goodacre on the floor near the east border.

Between the Mare Humorum and the south pole are light-colored rugged uplands with not many notable formations. The first we encounter are the two adjacent formations Hainzel and Mee. Hainzel is an interesting complex structure formed by partial coalescence of two rings of nearly equal size. The north–south dimension is about 60 miles and the breadth is not quite half of this. The walls are very high, the average height being 9000 feet. Hainzel should be observed at sunrise, when about half of the floor is illuminated, for the true structure to be revealed. The north component contains several craters and there is a curved mountain arm in the southern component. Hainzel almost disappears at full Moon. To the south-east of Hainzel is the large mountain walled plain with low walls named Mee. Further south is the elongated trough-like form of Schiller, an elliptical formation, which like Hainzel appears to have been formed by the fusion of two ring structures. There is hardly any floor detail apart from a few ridges and craterlets. To the south are the close crater pair Röst and Weigel, respectively 30 and 20 miles in diameter.

We now swing the telescope north again to the area just north of Gassendi, where we find the similar-sized but ruined ring of Letronne on the edge of the Oceanus Procellarum. Its walls are traceable on the southern half but the remainder are virtually gone. The whole formation gives the impression of being almost completely submerged below the Oceanus surface.

Fig. 4.33 Hippalus clefts. Redrawn after W. Goodacre. F.W. Price

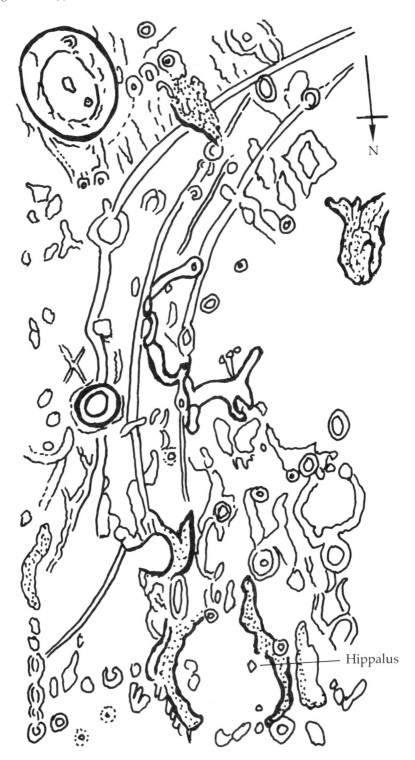

N

Hippalus

On its floor is a central peak that was seen as two shallow craters by A. Mee. T.G. Elger mentions a group of four bright mountains on the interior, three of which form a triangle. To the north again may be traced the faint outlines of a similar-sized ring on the Oceanus surface on the south rim of which stands the nine-mile crater Flamsteed. The large ring appears almost complete at full Moon and this with Letronne and Gassendi form a fine meridionally aligned group of formations.

On the Oceanus surface between here and the Sinus Iridum are some notable craters. Of these, Kepler is most prominent as it is bright and the center of an extensive ray system. It is 22 miles in diameter with terraced walls and a central hill. To the south is Encke, 20 miles in diameter with a floor depressed about 1000 feet below the surrounding Oceanus Procellarum. A high ridge runs along the floor from north to south and divides into two at its north end.

The Sinus Iridum is now seen in its entirety and some distance to the south of Cape Heraclides on the Oceanus Procellarum is the meridionally aligned pair of craters Diophantus and Delisle, respectively 13 miles and 16 miles in diameter. There are several low domes in their vicinity.

Among the Jura Mountains bordering the Sinus Iridum are some fair-sized craters among which are Bianchini almost on the edge of the bay and further away to the south-east Sharp and Mairan. Bianchini is polygonal and 25 miles in diameter. It has a central mountain on its floor. Sharp is 22 miles across, has a central mountain and there are indications of terracing on its inner east slope. The 25-mile irregular-shaped Mairan has lofty walls rising to 15 000 feet above the floor on the west. Schröter shows a central mountain in Mairan and is apparently the only selenographer to have done so. T.G. Elger reports seeing a low central hill near the center on several occasions but Wilkins and Moore express doubt about whether there is a central eminence in Mairan.

North of the Sinus Iridum on the Mare Frigoris is the deep 22-mile crater Harpalus with walls that tower to a height of 16 000 feet on the east and a bright central mounatin. Harpalus is the center of a minor ray system.

Further north in the light-colored upland area on the other (north) side of the Mare Frigoris is a very large formation 90 miles across which is essentially a portion of the surface bordered by ridges. It is named J. Herschel. On its vast interior are many craters, ridges and hills and in the center is the largest of the craters. At sunrise, the many longitudinal ridges lying close together on the interior have the appearance of fine scratches or grooves of variable breadth and depth.

Adjoining J. Herschel on the east is Anaximander, a splendid object 54 miles in diameter and seen in much foreshortened perspective. J. Schmidt shows a crater and some details on its floor. Encroaching on its north-west wall is Carpenter, a fine regular ring with a central mountain.

Between Carpenter and Philolaus is Anaximenes, a large 65-mile ring with a rather dark smooth floor. Its walls rise to a height of 8000 feet on the east. Four craters are shown on the west part of the floor by J. Schmidt and another on the south-east side. On the interior is a bright streak reaching southwards some way across the Mare Frigoris.

159

Twelfth day

This evening we are treated to the spectacle of sunrise on Schickard, one of the largest walled plains on the Moon. Schickard is in the southern hemisphere in light-colored upland terrain and is at about the same latitude as Mee. It is 134 miles in diameter but has only low walls, the greatest heights being peaks rising to 9500 feet on the west. The overall wall height

Day 12.

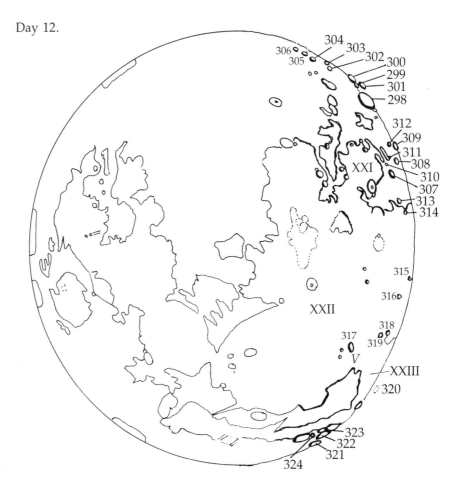

Day 12. Key to formations

298. Schickard	309. Vieta	320. Rümker
299. Nasmyth	310. Liebig	321. Pythagoras
300. Phocylides	311. De Gasparis	322. Babbage
301. Wargentin	312. Fourier	323. South
302. Segner	313. Billy	324. Robinson
303. Zuchius	314. Hansteen	
304. Bettinus	315. Reiner	
305. Kircher	316. Marius	XXI. Mare Humorum
306. Wilson	317. Aristarchus	XXII. Oceanus Procellarum
307. Mersenius	318. Herodotus	XXIII. Sinus Roris
308. Cavendish	319. Schröter's Valley	
		V. Harbinger Mountains

is 4500 feet. The interior is variegated by two large dark patches, a sharply defined triangular patch on the south-west (oddly, T.G. Elger describes it as oblong) and a dark area covering most of the north part of the floor. Through this in a north–south direction runs a light streak upon which is a row of small craters. On the interior are three large ring plains, all south of the center, and there are several smaller ones close to the foot of the inner east wall. The latter can only be seen well at favorable libration. Strange and inexplicable appearances have been seen several times within Schickard. On August 31, 1944, H.P. Wilkins was observing Schickard with an 8½-inch reflector and saw very few craters on the floor when several should have been visible. Instead, there were many light spots contrasting strongly with the dark areas. This appearance was seen by P.A. Moore twice in 1939. Mists have been suspected of obscuring details on the interior from time to time. These phenomena will be referred to again in Chapter 6. Almost adjoining Schickard on the south is a close group of formations, Nasmyth, Phocylides and Wargentin, but these will be better seen on the following evening as they lie a little too far beyond the terminator tonight. Still further south are a group of craters, all more or less of the same size as each other, forming an approximately north–south chain. From north to south these are: Segner, 45 miles in diameter on the south-east of Schiller. It has a small but conspicuous central mountain and a peak on the west wall casts a remarkable shadow. Adjoining Segner on the south-east is Zuchius (50 miles), the center of a minor ray system. There is a compound central mountain. A fine chain of craters runs around the outer west slope of the wall and these are best seen when the morning terminator is on the opposite border. W. Haas finds that at sunset the shadow within Zuchius is darker than those in other formations in the area. Bettinus, to the south of Zuchius, has lofty walls, some peaks rising to 13 000 feet. The inner walls are terraced and there is a central mountain, the summit of which catches the sunlight long before any part of the deep interior is lit up. Moving further south brings us to Kircher, another fine formation 45 miles in diameter with prodigiously high walls on the south rising to 18 000 feet above the depressed floor. T.G. Elger states that the floor appears to be devoid of detail; Lohrmann drew a central peak and Wilkins and Moore mention a ridge on the floor. The last of this series of five formations is Wilson, 40 miles in diameter, a grand crater with very high walls and again reported by T.G. Elger as having no conspicuous features on its level floor, but there is a crater on the inner slope on the west and a low ridge on the floor according to Wilkins and Moore.

If we now swing the telescope north again to the light-colored upland area east of the Mare Humorum, we will notice at once a cluster of craters, of which three are quite large and form a curving group. These are Mersenius, Cavendish and Vieta. Mersenius is the most northerly and lies close to the north-east border of the Mare Humorum. It is 45 miles in diameter and is noted mainly for its distinctly convex floor and the extensive rill system to the west. On the floor is a craterlet nearly at the center which is one of a chain extending from the south wall, and there are several clefts. The inner walls are terraced and rise to a height of 7000 feet on the west and are even higher on the north. The Mersenius rills all run roughly parallel to the north-east side of the Mare Humorum and reach to the Percy Mountains. One of these rills cuts through the ring structure called Mersenius D which lies to the south-west of Mersenius. Between Mersenius D and Vieta, forming a straight line with Mersenius D, are the craters Liebig (formerly Mersenius A) and De Gasparis (formerly Cavendish D), both somewhat larger than Mersenius D and all closely associated with the rill system in this area. The second of the trio of large formations is Cavendish, 32 miles in diameter with broad terraced walls and some low rings on its floor. The south-east wall is broken by a bright ring plain E, 12 miles in diameter. The third formation in this group is Vieta, 50 miles in diameter and considered by Elger to be one of the finest objects in the Moon's south-east quadrant. It has a small central hill and several craterlets on its floor; Wilkins and Moore mention six craterlets and a ruinous ring close to

Fig. 4.34 Chart of Schickard (from a photograph). F.W. Price

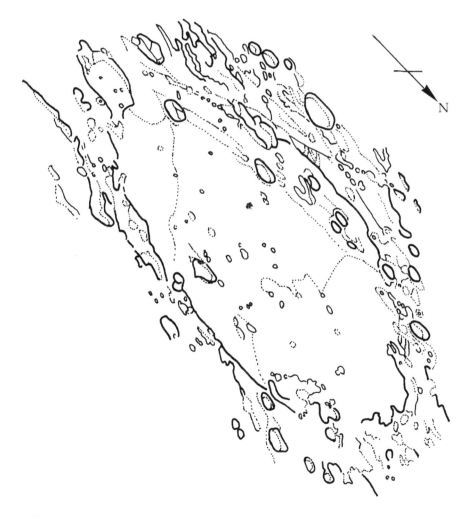

the north wall but Elger mentions ten craterlets on the north half of the interior. I have observed Vieta with an eight-inch refractor and have found seven craters on the north half of the floor. To the south-west of Vieta is the smaller (30 miles) ring of Fourier. It has terraced walls rising at one peak on the west to a height of 9500 feet above the floor and there is a small central crater. There are two craters on the outer slopes of the north-west wall.

Northwards from this cluster of craters and close to the south 'shore' of the Oceanus Procellarum are the two prominent craters Billy and Hansteen. Billy, the more southerly of the pair, is roughly south-east of Letronne. It is 31 miles in diameter and appears polygonal or elliptical in outline depending on the libration. The floor is flat and rather dark with two light spots on the northern part and two small craters in the south half of the interior. Hansteen is 32 miles in diameter with terraced walls rising to 3800 feet above the floor upon which are some hills, bright curved ridges and a craterlet near the north wall. On the Oceanus surface near to the south-west

Fig. 4.35 Mersenius. April 9, 1968. Colong. 56.0°. F.W. Price. Eight-inch refractor. Buffalo Museum of Science, NY, USA

border of Hansteen is a curious triangular mountain mass with a small bright crater upon it and blunt finger-like projections on the south.

For quite some considerable distance to the north of Hansteen there is nothing very much of note on this part of the vast expanse of the Oceanus Procellarum until we come to two isolated craters widely separated from one another. First is Reiner, 20 miles in diameter with bright terraced walls reaching a height of 10 000 feet above the floor, which is dark with two hills upon it. On the outside of Reiner a ridge runs from the north and south wall. To the east-north-east on the plain is a curious large white marking that resembles a 'Jew's harp' in shape. Some distance to the north-north-west is Marius, 26 miles in diameter, with a bright border rising to 4000 feet above its floor. There are many interesting objects on the interior.

Fig. 4.36 Vieta. February 21, 1986. Colong. 64° app. F.W. Price. Eight-inch reflector.

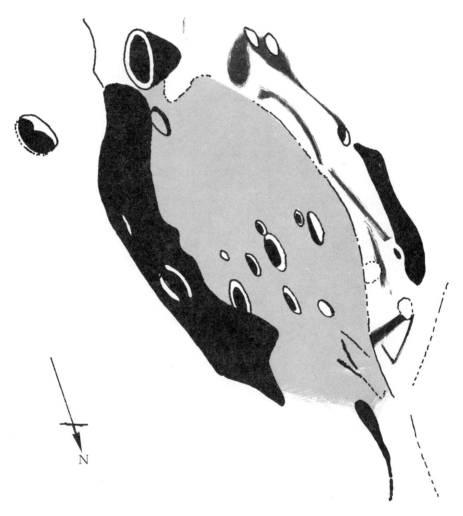

N

Fig. 4.37 *Aristarchus and Herodotus. February 27, 1980. Colong. 55.6°. F.W. Price. Eight-inch reflector.*

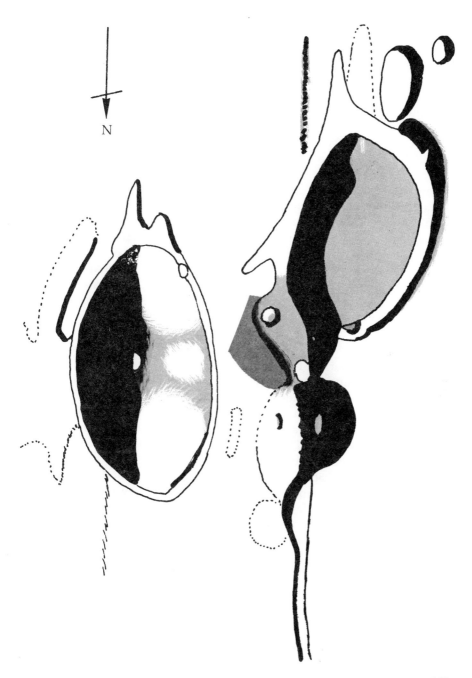

N

Beer and Mädler showed nothing here but Gruithuisen detected a crater close to the wall on the north-west and this was seen by Webb in 1864. Elger says that this crater stands on a light circular area and mentions three other white spots on the floor, one a little south of the center, a smaller one south-east of this and another near the inner foot of the south-west wall. There is a convex central eminence. On January 2, 1901, S. Bolton was observing Marius with a 4½-inch refractor and saw a low ridge running longitudinally and a dark triangular marking reaching from the central eminence to the east wall. These were subsequently confirmed by W. Goodacre on April 1, 1901. However, Wilkins was unable to find Bolton's ridge when observing Marius with the 33-inch Meudon refractor on April 7, 1952. Bolton also discovered a bright streak system on the interior that varies in brightness during different lunations. It is most prominent around the time of full Moon.

Again, for a considerable distance to the north of Marius there is nothing much to see until we come to the close pair of craters, Aristarchus and Herodotus, and the neighboring serpentine valley, which together are an arresting sight at lunar sunrise. Aristarchus is the brightest spot on the Moon and is the center of a system of bright rays that straggle over the surface of the Oceanus Procellarum. The white bowl of this 29-mile crater is quite dazzling at the full phase. It is easily visible in even a small telescope on the dark part of the Moon early in the lunation when the Earthshine is bright. As previously mentioned, Aristarchus was once so brilliant that Sir W. Herschel thought he was seeing an active volcano. The shape of Aristarchus is somewhat polygonal. No other lunar crater of this size has such prominently terraced walls, which are quite massive, and there are many prominent spurs and buttresses. Perhaps the most interesting feature of Aristarchus is the system of dusky interior radial bands that run from the floor up the inner walls. Two or three of these are distinctly visible in even a three-inch refractor but there are at least nine running up the north, east and south walls and presumably up the west wall, which is tilted away from us owing to the nearness of Aristarchus to the limb. Oddly, not much notice seems to have been taken of these features by the earlier selenographers, Mädler, Schmidt and Neison. Lord Rosse showed them distinctly in one of his drawings which was made five years before the time when Phillips is supposed to have first recorded them in 1868. They appear to have increased in visibility since they were first recorded. During each lunation they gradually increase in number and visibility as the sun climbs higher. Herodotus, 23 miles in diameter, contrasts strongly with its companion. It has a dark level floor and the only detail upon it is a light streak in the south part of the floor that appears under a high Sun. It crosses from wall to wall in an east–west direction. The walls of Herodotus are rather narrow and rise to about 4000 feet. Of principal interest is the long winding cleft, the first, of its kind to be discovered (in 1877), named Schröter's Valley after its discoverer, which begins inside Herodotus at the north wall; however, it needs a large telescope and the right lighting to be convinced of this. It starts as a valley which narrows to a fine cleft and this runs into a wide enclosure apparently formed by fusion of two craters. Because of its shape this was named the Cobra-Head by W.H. Steavenson. The valley then narrows and runs for a short distance in a northerly direction, takes a sudden bend to the north-east and continues for several miles before turning south-east and running for a similar distance. It then becomes a fine cleft and passes around the south end of a plateau and finishes at a mountain mass. The valley is bright throughout its course and the snake-like twists and bends and the Cobra-Head give it a striking resemblance to a snake rearing its head in response to the snake charmer's flute!

Just about on the terminator at this phase is the odd formation Rümker, which is situated north-east of Aristarchus and Herodotus on the extreme north Oceanus Procellarum due east of Cape Heraclides. It is a plateau about 30 miles across. The marial surface south of Rümker at the extreme east end of the Mare Frigoris is known as the Sinus Roris and in this region between Harpalus and the limb is a group of large walled plains, a brief study of which will conclude our survey of the 12-day-old Moon. The most impressive of these is Pythagoras, the nearest of the group to the limb. It is one of the largest of the lunar ring formations and is 75 miles in diameter with beautifully terraced walls with peaks rising to 17 000 feet in places and a splendid multiple central mountain group on the floor. It is a pity

Fig. 4.38 Dusky bands in Aristarchus. May 27, 1961. F.W. Price. Three-inch refractor.

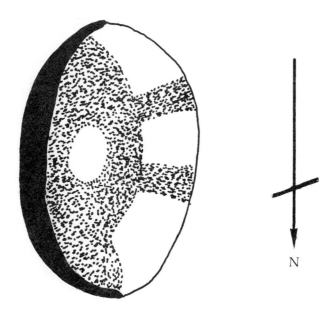

N

Fig. 4.39 Babbage at lunar sunrise. January 5, 1974. Colong. 56.7°. F.W. Price. Three-inch refractor.

that Pythagoras is not more centrally placed on the Moon's disc for it would be a truly imposing sight. Between Pythagoras and Harpalus are the walled plains Babbage and South which share a wall in common. Babbage is not really a crater and has been formed by fusion of two large rings. On the floor of the southern component, south of center, is the deep crater A. At lunar sunrise I have seen the rim of A brightly illuminated and standing out splendidly from the surrounding blackness like a silver ring. When the Sun is higher, the floor of Babbage is seen to be crossed with transverse ridges similar to those in Herschel but thicker. Babbage's companion South is only a little smaller and has low broken walls and a very rough interior with many ridges crossing it. There is no crater or mountain at the center. Close to South on the north is the deep 12-mile crater Robinson. It has a smooth floor, terraced walls and a craterlet on its west wall.

Thirteenth day

This evening, the terminator passes through another longitudinal chain of large formations that lie close to the east limb. The largest and most southerly of these is the great walled plain Grimaldi, 120 miles in diameter, with discontinuous walls averaging 4000 feet in height. On the inner north-east wall is the bright crater Saheki, formerly Grimaldi B. Grimaldi is so large and has such a dark floor that if it was more centrally placed it would most certainly have been classified as a minor Mare. At colongitude 132–140 degrees, W. Haas noticed a greenish tinge to the floor, confirmed by Pickering; Wilkins observed dark spots on the interior that he described as 'variable'. The lunar observer should not miss any opportunity to watch the Sun rise on Grimaldi; it is one of the most spectacular sights on the Moon. A little way to the north is the much smaller formation Lohrmann, 28 miles in diameter. It has a bright wall and a rather dark floor with a central hill. Several clefts run from the outside walls in different directions. North again is Hevelius, 70 miles across with terraced inner slopes and walls with peaks rising to 6000 feet on the west. There are three prominent craterlets on the inner slope of the east wall. The floor is convex, has a low triangular central hill and there are several clefts. Under early morning illumination, three remarkable light streaks or ridges are seen crossing the north-east part of the floor, well shown in a drawing by S.R.B. Cooke, dated September 10, 1935. Cavalerius, the most northerly member of the chain, is 40 miles in diameter and has terrraced walls rising to 10 000 feet above the floor. The central mountain is elongated and has three peaks.

To the west of Grimaldi is Damoiseau, a formation that is especially interesting and almost unique among lunar objects. It is a complex of rings and ridges, the largest being 23 miles in diameter inside of which is a smaller eccentrically placed ring. Within this is a curved ridge and some craterlets. There is a still smaller ring to the west of Damoiseau.

If we now sweep in a northerly direction along the terminator, passing the little crater Galilei, there is nothing much of note until we reach the isolated craters Seleucus and, further north, Briggs. They lie eastwards from Aristarchus and Herodotus and exactly between these and the other

Day 13.

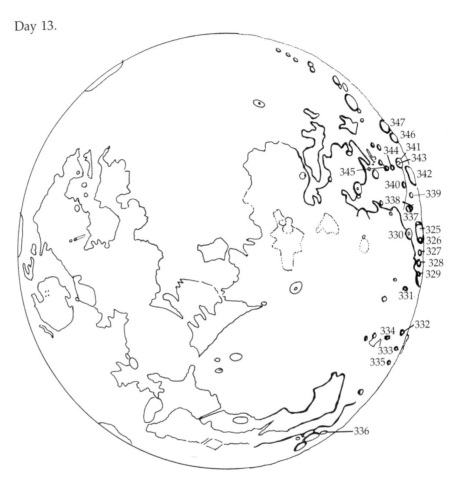

Day 13. Key to formations

325. Grimaldi	337. Sirsalis
326. Saheki	338. Bertaud
327. Lohrmann	339. Crüger
328. Hevelius	340. De Vico
329. Cavalerius	341. Byrgius
330. Damoiseau	342. Darwin
331. Galilei	343. La Paz
332. Seleucus	344. Henry, Prosper
333. Briggs	345. Henry, Paul
334. Schiaparelli	346. Lagrange
335. Lichtenberg	347. Piazzi
336. Oenopides	

Fig. 4.40 Clefts in Hevelius. November 9, 1981. Colong. 68.9°. F.W. Price. Eight-inch reflector.

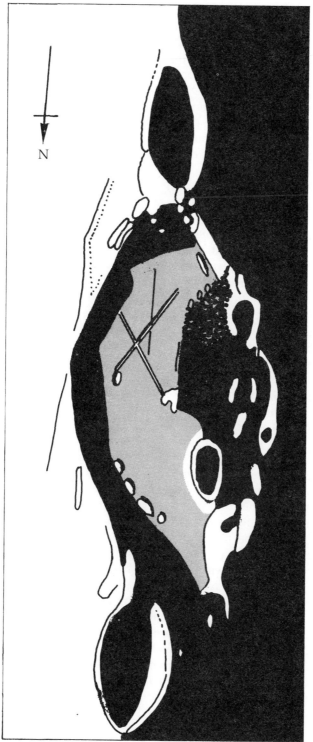

two craters is one called Schiaparelli. Seleucus, 32 miles in diameter, is a fine specimen of a crater. Its terraced walls attain a height of 10 000 feet above its very deep dark floor upon which is an inconspicuous central mountain. There are dusky bands on the interior slopes according to A.P. Lenham but P. Moore was unable to see them with either a 12½-inch reflector or the 33-inch Meudon refractor. Briggs is a crater 33 miles across with a level floor upon which is a long ridge running in a north–south direction. To the north is a crater Briggs B.

Further north again is another smaller isolated crater Lichtenberg, 12 miles in diameter. It is the center of a minor ray system. Under high light it is a diffuse whitish nimbus. Over the years there have been reports of a reddish color on the surface closely west of Lichtenberg under certain illumination angles.

Much further north and impinging on the south-east wall of Babbage is Oenopides, 42 miles across with rather irregular walls. Elger says that the floor is apparently devoid of detail but Wilkins and Moore mention two craterlets, a hill and two pits. There are some depressions on the west wall that are worth looking at at sunrise.

Returning now to Damoiseau and scanning southwards along the terminator we soon encounter Sirsalis, diameter 20 miles, the more westerly member of a formation consisting of two rings pressed close together, the other being named Bertaud. It lies in the rugged mountainous area south-west from Grimaldi and has a diameter of 20 miles. Immediately to the west of Sirsalis runs one of the longest clefts on the Moon's Earthward face. It commences at a minute crater to the north-west of Sirsalis, runs south-east to skirt the floor of the west wall and immediately afterwards passes between a close pair of craters much like a smaller version of the Sirsalis–Bertaud pair. Still running in a south-east direction, the cleft passes the crater Crüger A on the west and then turns slightly to the south and connects with De Vico A, a ring plain east of De Vico, which it traverses, and continues onwards changing direction slightly and finally enters the interior of the crater Byrgius. There are several branches in the cleft at various places along its course and much of it consists of rows of craterlets. The principal cleft is an easy object but the complete system including the subsidiary clefts is not often seen as it needs favorable conditions and a good telescope.

At this phase the terminator will be just grazing the west wall of the large formation Darwin, which is north-east of Byrgius, but it will be best to defer study of Darwin until tomorrow evening. Byrgius is a crater 40 miles in diameter and has walls rising to 7000 feet. On its west crest is a very bright crater, La Paz, formerly known as Byrgius A. It is the center of a ray system, most of the rays extending eastwards. One on the west reaches to Cavendish and another to Mersenius. Between Byrgius and Mersenius are the two large rings Henry, Freres – Paul and Prosper. South of Byrgius are two large walled Plains, Lagrange and Piazzi. Lagrange is over 100 miles across with a more or less complete wall which has terraces on the east. This is a fine telescopic sight at sunrise when the libration is favorable. On the interior is a meridionally oriented central ridge and a crater chain, more craters and a dark spot on the south-east.

Fig. 4.41 Damoiseaux. April 9–10, 1979. Colong. 73° app. F.W. Price. Eight-inch reflector.

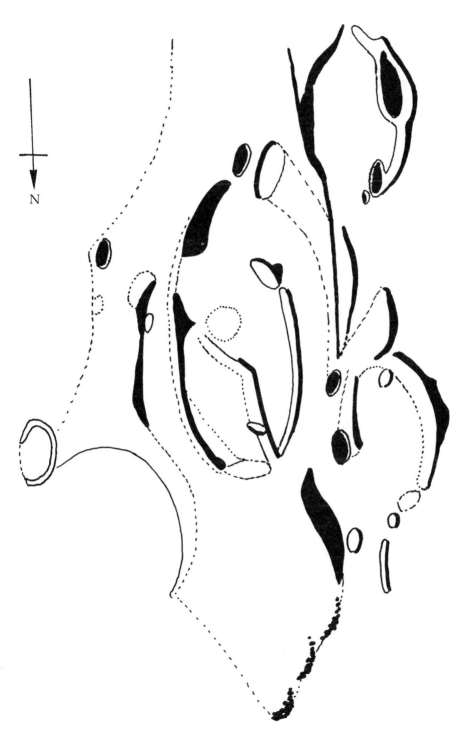

Similar to Lagrange and not quite as large is Piazzi, which lies immediately north-west of Lagrange. It is 80 miles in diameter and has complex broken walls of irregular height. There are several depressions on the north-west and on the dark floor is a prominent central mountain, mentioned by Elger, but Wilkins and Moore omit mention of it although they enumerate other interior details.

Some distance to the south and closely south of Schickard is the earlier mentioned group of formations, Nasmyth, Phocylides and Wargentin. Phocylides, and most southerly of the three, is a crater 60 miles in diameter. Nasmyth adjoins it on the south, the two being separated by a common wall which appears to be a high cliff or fault whose shadow under low Sun has a striking appearance. The floor of Nasmyth is 1500 feet higher than

Fig. 4.42 Phocylides, Nasmyth and Wargentin. (Based on 36-inch telescope photograph). F.W. Price

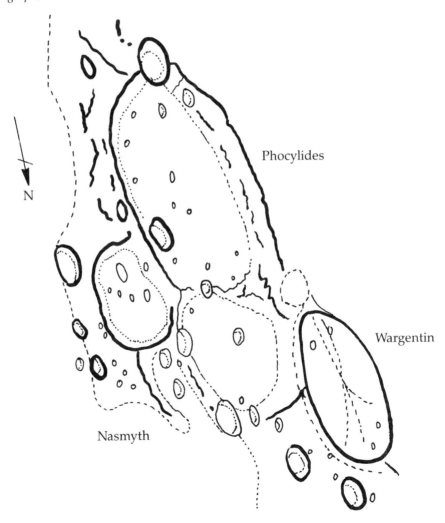

that of the floor of Phocylides. These two formations together have an uncanny resemblance to a shoe print, Phocylides forming the 'sole' and Nasmyth the 'heel'. On the floors of both these formations are many craterlets. Haas reports that the sunrise shadow in Phocylides is darker than that within Schickard. The third formation of the group, Wargentin (55 miles in diameter), adjoins Nasmyth on the north-east and is probably almost unique among lunar formations in that contrary to the usual situation its floor is raised above the level of the surrounding country. It looks as if the interior of Wargentin became filled with molten lava almost to the level of the top of its walls at some remote time. Its appearance has been likened to an oval slab of cheese. There is little to be seen on the raised interior of Wargentin with moderate telescopes except a branching ridge radiating from near the center·that looks like a bird's foot. Wilkins reports two dark markings on the north part of the floor and R. Barker believes that these are 'variable'. With this, we conclude our study of the 13-day-old Moon.

Fourteenth day

The large formation Darwin that we noted last night is now well exposed for observation. It lies to the north-east of Byrgius and is seen foreshortened into a long ellipse owing to its closeness to the limb. It is a vast depression walled in by mountains that are rugged but not especially high. On its rather dark floor are interesting details including the largest dome on the Moon's Earthward hemisphere. It was discovered by R. Barker and is visible in even a three-inch refractor although its form is better appreciated with larger apertures. The dome is traversed by two cracks one of them a branch of the Sirsalis system.

Between Darwin and the south part of the terminator are some other large formations. To the south-east of Wargentin is the ring plain Inghirami, 60 miles in diameter, with a depressed interior and walls rising to 12 500 feet above it, which are beautifully terraced. There is a small mountain and craterlets on the floor. On the inner east wall is a group of dark band-like features which were discovered by P.A. Moore and others using a 9½-inch reflector on August 23, 1953. There are two curious dark spots on the north part of the interior; these are probably depressions. Right on the terminator and about half way between Inghirami and the south pole is the largest of all the lunar walled plains, Bailly, the maria only being excepted. It is 183 miles in diameter and has peaks on its walls attaining heights of 10 000 to 13 000 feet on the west and 14 000 feet on the east. The east wall is beautifully terraced. On its huge floor are ridges and some craters. On the south-east are many remarkable parallel curved valleys traversing the border. On the east side of the floor are two fine dark lines, probably clefts, that cross each other near the south end. The largest crater on the interior is Hare, near the south wall, which abuts on a smaller crater, A, on the south wall. In order to observe Bailly satisfactorily, the libration has to be favorable.

South of Bailly and very close to the limb is Legentil, a large crater only seen well when the libration is favorable. It has high walls and some craters on its floor. To the south-west is the large ruinous formation Drygalski.

If the telescope is now swung back 'down' the terminator to the region immediately north-east of Grimaldi, we will see the Sun rising on the large formation Riccioli that lies close to the north-east wall of Grimaldi. It is 100 miles in diameter and has rather low broken and discontinuous walls. The interior contains much interesting detail, most of which was discovered by H.P. Wilkins with the 33-inch Meudon refractor. Nearly at the center is a high hill standing on its own. Of the other floor detail, moderate apertures reveal a few low rounded hills and ridges, two small craters on the south-west and a short row of craters under the south-east wall. The floor itself

Day 14

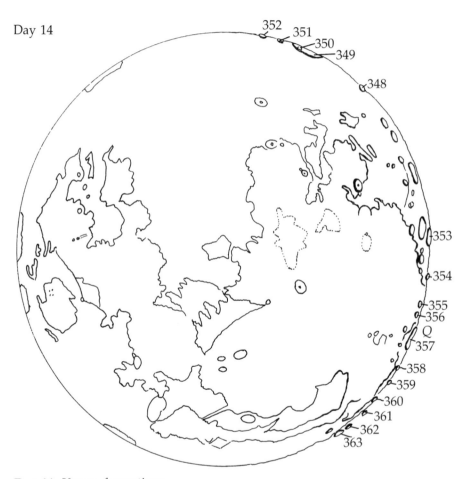

Day 14. Key to formations

348. Inghirami	357. Otto Struve
349. Bailly	358. Ulugh Beigh
350. Hare	359. Lavoisier
351. Legentil	360. Gerard
352. Drygalski	361. Galvani
353. Riccioli	362. Regnault
354. Olbers	363. Xenophanes
355. Cardanus	
356. Krafft	Q. Hercynian Mountains

is remarkable for its duskiness, especially on the north. The west side of the dusky area is especially dark, apparently even darker than any part of the floor of Grimaldi. Many lunar observers say that this dark patch changes shape during a lunar eclipse. The north part of the wall of Riccioli is traversed by several ridges with valleys in between and the south wall is cut through by a valley.

Further 'down' the terminator and north-east of Cavalerius is Olbers, a crater 40 miles in diameter, which is the center of a ray system. The rays lie mostly to the east and are not easy to see.

North from Olbers is a somewhat widely separated meridionally arranged pair of craters, Cardanus and Krafft. Cardanus, the southern member of the pair, is a prominent ring 32 miles in diameter with a central mountain and several craterlets on the interior. It has bright terraced walls rising to 4000 feet above the floor, which is light grey in color. There is a fine deep valley on the outer west slope. Krafft, 30 miles in diameter, has a central mountain and a large crater on its dark floor close against the south-west wall. On the outer side of the west border is a smaller crater from which a cleft bordered by a bright band stretches to the north wall of Cardanus. This is best seen when the terminator coincides with the east wall of Cardanus.

North of Krafft and even closer to the east limb is the large formation Otto Struve, apparently the result of fusion of two ring structures each about 100 miles across and oriented in a north–south direction. The wall that once divided them has entirely disappeared. The interior is so vast that the convexity of the Moon's surface is clearly revealed. On the west of the uneven-toned floor are four craters, and many hills and light spots. Two whitish ridges extend from the center to the south wall and there is a craterlet near the north wall. Abutting on Otto Struve to the west is the smaller enclosure Otto Struve A, which is bordered by ridges. When the Sun is rising on this formation the east and west walls with the mountain mass at the north end joining them looks like a lobster's claw or a pair of partly open calipers. East of Otto Struve are the Hercynian Mountains.

Under conditions of favorable libration a few formations north of Otto Struve may be visible. There is Ulugh Beigh, 30 miles in diameter, with narrow walls and a central mountain which is the highest point of a long ridge. Progressively further north are the craters Lavoisier (40 miles), Gerard (50 miles), Galvani, Regnault and Xenophanes, the last of which is 67 miles across. However, like Ulugh Beigh, all are really too close to the limb for easy observation.

Fifteenth day – full Moon

The full Moon is almost dazzling when viewed in even an eight-inch telescope, especially if a low power is used. Gone now are the dramatic high contrast views we have been enjoying of lunar features when they were on or near the terminator. Even their most delicate details were boldly revealed by the strong shadows they cast. All that we see tonight is a patchwork of light and dark areas; contrast is minimal and many of the formations that were so prominent and imposing on previous evenings are now difficult or impossible to locate. Everything is engulfed in the general glare because the Sun's light is now being reflected directly into our eyes. Many craters with bright rims will now be seen as brilliant white circles against the dusky background and their central mountains and inner craterlets where present will often appear as white spots. In many cases, crater floors that earlier appeared to be evenly tinted will now be variegated often with patterns of white spots and patches whose presence was previously unsuspected. The 'variable' dark spots within craters like Atlas and Alphonsus are now fully 'developed'. In fact, the most noticeable thing about the

Day 15.

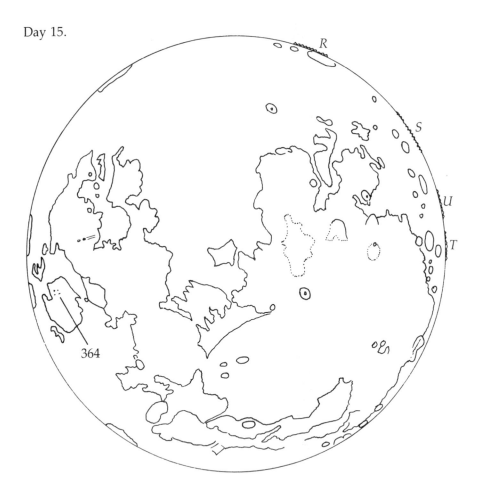

Day 15. Key to formations

364. Trapezium

R. Doerfel Mountains
S. Rook Mountains
T. D'Alembert Mountains
U. Cordillera Mountains

full Moon is the abundance of bright spots and streaks, especially the great ray systems of Tycho and Copernicus, which are the dominant features of the full Moon and stand out brilliantly against everything else. The most extensive is the ray system of Tycho. One of its rays appears to pass right across the Moon's disc crossing light and dark areas and then over the dark surface of the Mare Serenitatis. It can even be traced to the limb. At one time it was thought that this ray did, in fact, extend all the way from Tycho but more careful studies have revealed that the ray on the Mare Serenitatis

Fig. 4.43 Mountain peaks on the Moon's east limb. November 30, 1982. Colong. 94.4°. F.W. Price. Eight-inch reflector.

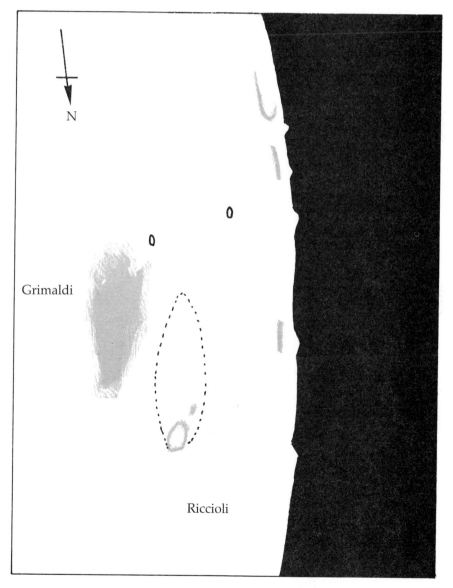

really originates from the crater Menelaus. The ray from Tycho ends at the border of the Mare Serenitatis.

Strictly speaking, we never see a really full Moon. For this to occur the Moon must be exactly opposite the Sun on the other side of the Earth, the centers of the bodies being in an exact straight line. However, under these conditions the Moon would be eclipsed by the Earth's shadow. The full Moon as ordinarily understood occurs when the moon is nearly, but not quite, exactly opposite the Sun. Therefore, the fullest possible Moon without an eclipse occurring will always show a slight phase effect at one or other of the poles. Because of this the full Moon phase is a good time to observe the polar regions, especially the south pole, where there is the very high Leibnitz Mountain range. Up until now we have had 'bird's eye' views of

Fig. 4.44 Mare Crisium – white spots and streaks, 1964. F.W. Price. Eight-inch refractor. Buffalo Museum of Science, NY, USA

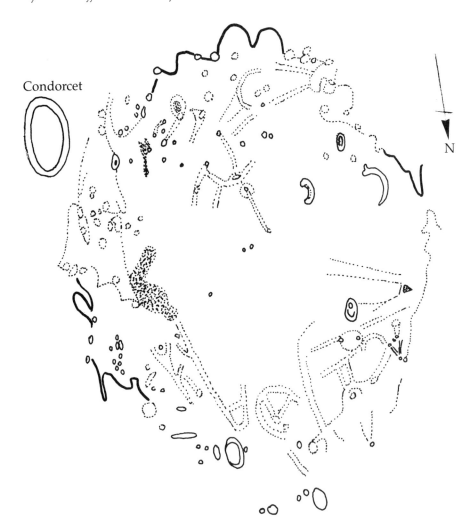

Condorcet

N

surface features in which we are looking 'down' on formations and seeing them in more or less plan view. At the poles and other parts of the limb, however, the mountains and other formations appear in profile and we see them as we would a landscape on Earth.

The east wall of Bailly continues into a very lofty mountain range, the Doerfel Mountains. The peaks of the Doerfels are some of the highest on the Moon, many towering to 26 000 feet. This range can be followed southwards where they merge into the even higher Leibnitz Mountains. When the libration is right the peaks of the Leibnitz range can be seen standing out from the Moon's limb against the black sky.

Beyond Schickard are the Rook Mountains, many of whose peaks rise to 20 000 feet at least. Due east of Riccioli, right on the Moon's east limb, are the peaks of the D'Alembert Mountains. The Moon's south and east limbs are therefore notable for their high mountain ranges. Also, to the east of Riccioli is a table-like mountain that has a flat top and sloping sides. This formation is brilliantly white and so it is frequently a striking object when seen against the blackness of the sky. Sweeping over to the west limb we come again to the Mare Crisium. It is often seen covered all over with white spots and streaks at full Moon. On the south part of the mare is a prominent 'constellation' of white spots known as the Trapezium or Barker's Quadrangle (named after the observer who drew attention to it many years ago). Curiously, the Trapezium was omitted from most of the older maps or only very incompletely delineated, yet it is very much obvious today in even a three-inch refractor. There has been discussion as to whether the Trapezium was simply omitted from the older maps or whether it has become more prominent since telescopic observation of the Moon began.

After full Moon

A little less than two weeks ago we commenced our nightly study of the Moon beginning with the three-day crescent Moon. The earlier phases of up to two days after new Moon, we noted, were not very satisfactory for observation owing mainly to the closeness of the Moon to the Sun and its consequent low altitude in the sky once it was dark enough to commence observing. This is why only the largest and most prominent formations were described. The best time to observe the formations close to the west limb is up to two days past full Moon. These are especially good phases to observe when it is winter as the Moon when even two days past full climbs high in the sky well before midnight and the seeing is apt to be best at this time of year. Of course, the Sun's light will now be falling on the formations from the opposite direction from what it was in the waxing crescent because the Sun is now setting on them. It is also a good time to observe the formations we saw in the three-day crescent under opposite lighting conditions, such as the Furnerius–Petavius–Vendelinus–Langrenus chain of walled plains.

Those formations that lie practically on the limb are seen in extreme foreshortening but some of them are worth looking at nonetheless when libration is favorable. One reason why formations very near the limb are difficult to observe, in addition to the foreshortening, is because as well as good seeing and the right phase of the Moon being necessary, libration also has to be favorable. This third variable does not apply, of course, to formations near the center of the visible disc.

Immediately after full, just as the west limb is showing a slight phase effect, the following formations may be conveniently studied. First, the

After full Moon.

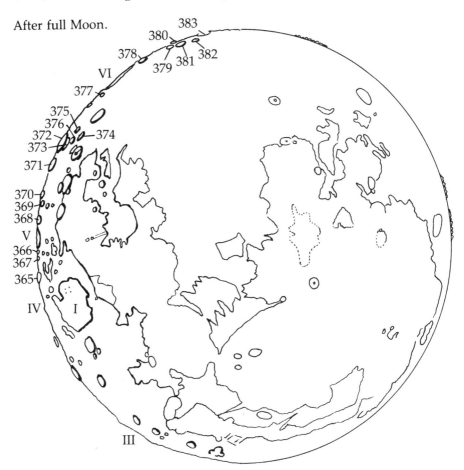

Afer Full Moon Key to formations

365. Neper	378. Pontecoulant
366. Febrer	379. Helmholtz
367. Schubert	380. Neumayer
368. Kästner	381. Boussingault
369. Lapeyrouse	382. Boguslawsky
370. Ansgarius	383. Demonax
371. Hecataeus	
372. Humboldt	I. Mare Crisium
373. Phillips	III. Mare Humboldtianum
374. Hase	IV. Mare Marginus
375. Adams	V. Mare Smythii
376. Legendre	VI. Mare Australe
377. Oken	

Fig. 4.45 Neper and north part of Mare Smythii. February 27, 1964. Colong. 275°
app. F.W. Price. Eight-inch refractor. Buffalo Museum of Science, NY, USA

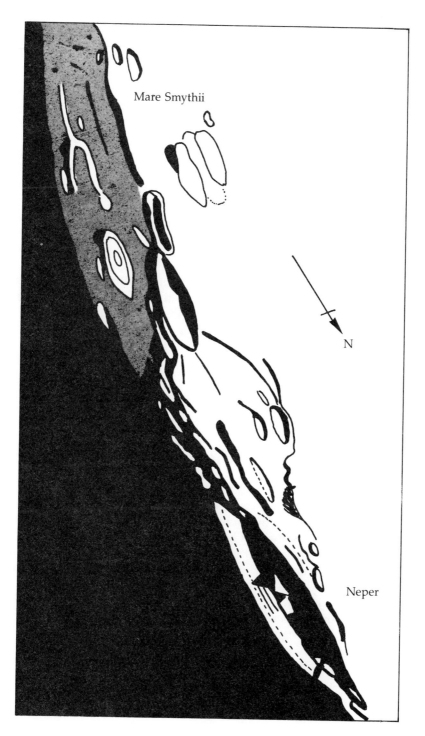

great walled formation Neper is worth locating. To find it, concentrate on the terminator directly opposite the south part of the Mare Crisium; very slightly south of this point Neper should be found without any difficulty. Indeed one can hardly miss it as it is 70 miles in diameter, with lofty terraced walls and a fine central hill which is the highest peak of a range of hills that traverses the interior. Neper appears as an elongated oval owing to extreme foreshortening; in fact its west wall sometimes actually lies on the limb. Because of this the inner surface of the west wall is well placed for observation as it faces straight at us. An eight-inch reflector reveals two elongated dusky lines on the wall concentric with the rim, which are probably related to the interior terracing.

On moving southwards along the terminator we come to the Mare Smythii which has interesting detail on its surface, and close by it are the formations Febrer and Schubert. Near the south end of Mare Smythii is the large enclosure Kastner, which has a smooth floor except for the ridge extending from the center to the north wall and two low peaks on the inner south that were seen by P.A. Moore on August 9, 1949, using a six-inch refractor.

South-west of Langrenus are Lapeyrouse and Ansgarius. Lapeyrouse, 41 miles in diameter, has a ridge on the center of its floor and in the southern part is a crater. Ansgarius, 50 miles across, is even closer to the limb. It has a few hills on its otherwise smooth floor. Southwards from Ansgarius is Necataeus, a pear-shaped formation with walls rising to 16 000 feet on the east and nearly the same height on the west. It appears to have been formed by fusion of two rings. There is a central mountain on the floor, some ridges and small craters on the south part.

To the west of Petavius are the two large ring structures Humboldt and Phillips, which share a wall in common. Humboldt, which is practically on the limb, is 120 miles in diameter and has lofty walls. In many ways it resembles Bailly. There is a chain of hills on the interior, a nearly central crater and two dark spots, one at the north and the other at the south end. Its companion Phillips, which impinges on its east wall, is 75 miles in diameter. There are several hills and ridges on the floor of Phillips including a long central ridge.

Roughly between Phillips and Petavius is Legendre, a 46-mile ring formation with low walls. There is a discontinuous longitudinal ridge on the floor with craterlets on either side of it. To the north of Legendre is Adams, which has a central crater and hills on its floor. Wedged between the walls of Petavius and Palitzsch on the south of these two formations is Hase, an irregular formation about 50 miles across. A large crater and three smaller ones are shown on the west side of the floor by Schmidt. In 1952, P.A. Moore noted eight craters on the west wall using his 12½-inch reflector. South-west from Furnerius and lying close to the limb is Oken, 50 miles in diameter with irregular broken walls. The floor has a rather dark tint and there are many ridges on it. On the inner east slope is a crater. The Mare Australe, a dark plain upon which are craters, ridges and ruined rings, extends along the limb from Oken to a fine ring formation close to the limb called Pontecoulant, which is 60 miles in diameter. It has a dark floor upon which are some craters and hills.

Fig. 4.46 Wilhelm Humboldt. February 27, 1983. Colong. 95.9°. F.W. Price. Eight-inch reflector.

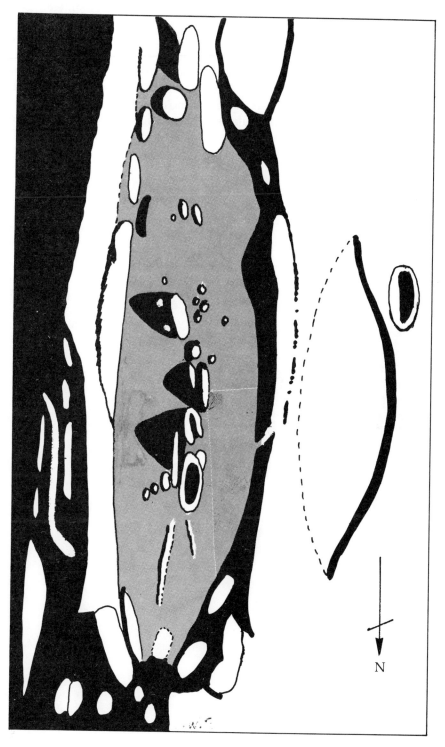

Some distance south of Pontecoulant is a group of large ring formations, Helmholtz, Neumeyer, Boussingault, Boguslawsky and Demonax. Helmholtz and Neumeyer, respectively 60 miles and 50 miles in diameter, form a close pair. Helmholtz has high walls and two mountains near the center of the floor. Neumeyer is very deep and closer to the limb than Helmholtz. Apart from two hills on its north-west portion, the floor is smooth. P.A. Moore has noted some curious ray-like features traversing Helmholtz and Neumeyer, apparently originating from the Moon's far side. Boussingault, to the east of Neumeyer consists of three almost concentric rings, the outer one being 70 miles across. Boguslawsky, due south of Boussingault, is 60 miles in diameter with terraced walls and on the floor is a large crater. To

Fig. 4.47 The Cape Agarum at lunar sunrise. January 18, 1965. Colong. 165° app. F.W. Price. Eight-inch refractor. Buffalo Museum of Science, NY, USA

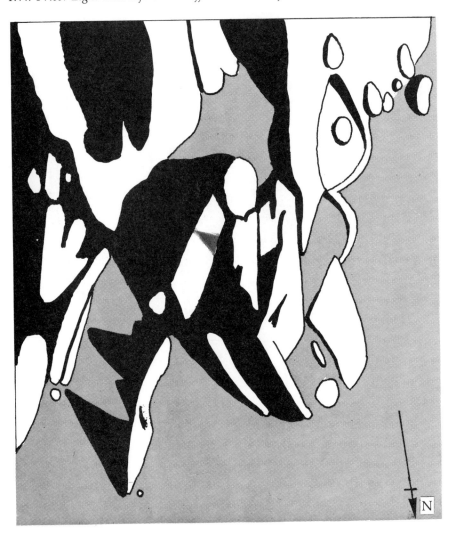

the west and nearly on the limb is Demonax, a great crater 75 miles in diameter, with two craterlets on its floor near the center and a ridge to the west. The Mare Marginis occupies that part of the limb opposite the south half of the Mare Crisium. Its surface has a rather light tint and on it can be seen many large ruinous ring structures.

The Mare Humboldtianum contains much detail including a long meridional ridge in the center and three domes on the west.

Other formations that may be studied immediately after full are those already mentioned under the three-day-old Moon. Up to two days after full is a good time for observing all these formations. Gauss is especially worth looking at. Perhaps the most spectacular sight of all is sunset on the Mare Crisium, when the evening terminator is anywhere between the west shore of the Mare and the east border of the Cape Agarum. The whole Mare really looks as if it were chiselled out of solid rock. The wrinkle ridge running longitudinally down its center and the diamond-shaped enclosure at its south end are prominent and the two northern brightest white spots of the Trapezium are clearly seen as shadow-filled craterlets, each casting tiny shadows on the mare surface. The whole Mare Crisium spectacle once seen is not easily forgotten and is definitely the show-piece of the Moon a day or two after full.

Further Reading

Books

Our Moon. Wilkins, H.P. Frederick Muller Ltd, London, England (1954). Chapters 4–9.

Astronomy for Everyman. Davidson, M. (ed.). J.M. Dent and Sons Ltd, London, England (1954). Chapter 3.

5

Observing and Recording

Lunar observations are recorded by drawing, making notes or both. Optimum conditions for observing lunar features depend on at least three factors:

1. The position of the terminator.
2. The state of the Earth's atmosphere.
3. The angular height of the Moon above the horizon.

The position of the terminator

The most dramatic and detailed views of lunar features are obtained when they are on or near to the terminator. Under these conditions, illumination is at a very low angle and long shadows are cast so that everything is seen in almost startling relief and contrast. If we are watching Sunrise the terminator moves away from a feature, the angle of illumination increases and the shadows shorten, drawing back and revealing detail that was previously hidden, but at the same time contrast is decreasing. Finally, when the Sun is almost overhead, shadows are shortest and contrast is minimal. At full Moon, Sunlight is reflected directly into the observer's eyes and everything is dazzling and bright. (Although the Sun is directly overhead only at a point near the Moon's apparent center at full Moon, we see no shadows over the entire disc because they would lie on the side away from us of the objects casting them.) Rugged surface detail previously seen near the terminator quite readily and clearly defined may now appear diffuse or will seem to have disappeared. At the same time light or dark spots, patches or streaks that were previously invisible under low angle illumination may make their appearance. An example is provided by the bright rays surrounding certain craters. These are at their brightest when the Moon is full but may be invisible when near the terminator.

The state of the Earth's atmosphere

Unfortunately for us, we have to study the Moon from the bottom of a vast ocean of air which is rarely, if ever, quite still. Temperature differences and air currents at different levels of the upper atmosphere all combine to cause irregular refraction of the light rays reaching us from the Moon and this adversely affects the quality of the 'seeing'. The effect is to make the telescopic image tremble or shimmer to a greater or lesser degree and difficulty is thereby experienced in trying to observe the smaller and more delicate features on the lunar surface. There is no cure for this except to wait until conditions improve. Generally, the best seeing occurs on winter nights when the temperature is close to or below freezing, the sky clear and the air still. The best telescopic views of the Moon I have ever enjoyed have been on such nights. On these occasions the seeing can be astoundingly good. Such excellent seeing I call 'crystalline' and the amount of detail 'brought out' by the telescope is quite breathtaking. Under these conditions the optical excellence of a good telescope mirror or lens is truly revealed.

When the seeing is unsteady – as it usually is – there is no need to despair; brief steady intervals occur, often quite frequently, if the unsteadiness is not too great. Part of the art of being a good observer is learning to wait for and make the best of these brief steady intervals. In my experience, however, most of one's patience and nervous energy is sapped by this watching and waiting for steady intervals. You will only fully realize this when you encounter one of those rare nights of 'crystalline' seeing when all of your time and energy can be devoted to observing and drawing; at these happy times the feature you are delineating almost 'draws itself'.

When the transparency of the air is not quite perfect, as when there is some slight overhead mist, the air is often quite steady and the seeing good.

The worst seeing occurs in summer, one cause being the heat given off by buildings or the concrete in your back yard at sunset and after, which causes local turbulence and unsteadiness of the air. Conditions often improve after midnight. Also, at this time of year, the Moon, especially around the full phase, is quite low in the sky where seeing is usually poor anyway.

The angular height of the Moon above the horizon

The higher the moon is above the horizon the steadier its telescopic image is likely to be. This is because the thickness of the layer of atmosphere through which the Moon is seen diminishes rapidly as the Moon's angular elevation above the horizon increases. On any evening, the Moon will be highest in the sky when it is on the meridian. During a single lunation this height can vary considerably, depending on the Moon's phase. During the year the height of the Moon when on the meridian at a given phase also varies considerably. The reason for these effects is explained in Chapter 2. On any given evening then, whatever the time of year, the Moon is best observed when on or close to the meridian, but there are times of the year that are best for observing particular phases of the Moon. At these times the Moon is at its highest when on the meridian at these particular phases than at any other time of the year. The crescent Moon when on the meridian, either three to four days or 25 to 26 days after new Moon, will be too close to the Sun to be easily visible as it will be broad daylight. At these phases, the Moon can only be observed after sunset (when three to four days after new) or before Sunrise (when 25 to 26 days after new). The Moon's height above the horizon will depend on the steepness of the ecliptic at sunset or sunrise and since this varies at different times of the year, so also will the Moon's height above the horizon at the crescent phases.

The table below shows the most and least favorable times of year in the Northern Hemisphere to observe the Moon at various phases. For the Southern Hemisphere interchange July with January and April with October.

Moon's phase	3–4 day crescent	First quarter	Full	Last quarter	25–26 day crescent
Most favorable	End of April	Spring Equinox	Winter Solstice	Autumnal Equinox	End of July
Least favorable	End of October	Autumn Equinox	Summer Solstice	Spring Equinox	End of January

Observing and drawing

An observation is useless if not recorded in some way, usually by drawing. Many beginners say 'I can't draw, I'm not an artist'. However, one does not have to be an artist because an artist creates whereas all we are doing in lunar drawing is copying what we see. Naturally, some people are gifted with the ability to draw but anyone, even if not talented, can learn to draw creditably if they are willing to try and to practice and not be put off by initial failures.

Of all the types of lunar features, the ring structures and craters are most likely to attract the attention of the lunar observer, especially if he is a beginner. In choosing a particular crater for study on a given night, with intentions of making a complete drawing, you should choose one of suitable size and detail. If the crater is too large and detailed then the production of a finished drawing showing all the fine structure that the telescope is capable of revealing may be physically impossible at one sitting. Most observers will not feel like enduring more than about one and a half hours of the cold of a clear winter evening. Even with a three inch refractor, a large crater like Clavius presents quite a challenge, so detailed is its structure. On the other hand, Plato, though still a large formation, is easier to handle at one sitting with such a telescope.

An important point when beginning a lunar observation is that you should be physically – and mentally – comfortable. On cold nights you should be warmly clothed. Have the telescope eyepiece at a comfortable angle and sit down to observe if you wish. One's physical state should be good; it is unwise to attempt serious observation if you are fatigued, hungry or have a streaming cold. One's frame of mind should be tranquil; personal cares and problems are best not brought to the telescope and should be dealt with before embarking on an observing program. If you find yourself becoming irritated because of poor seeing conditions or your drawing does not seem to be 'going right', it is best to stop and try again on another night.

Quite small craters (10–20 miles in diameter) are good to start with if you observe with a six–eight inch telescope and are good for practice before passing on to larger and more complex features.

The beginner should choose a crater under oblique illumination so that there is some shadow in its interior. Having decided on the crater you want to study and draw on a particular evening, first observe it carefully for several minutes. A good plan is to ask yourself questions about it; what is its exact shape–is it circular, polygonal or irregular in outline? Is the interior bowl-shaped or is there a flat floor? If flat is the floor evenly shaded or is it patchy? Is the floor seemingly flat or are there irregularities? Is there a central mountain or mountains? How many? Are there any craterlets on the interior of the crater? Any rills or cracks? Are the inner walls relatively smooth or are they terraced? Is the crater rim bright and if so is it equally bright all round? What are the exact shapes of any shadows? Are there any bright points visible within the shadows (elevations or protuberances just catching the Sun's light)? It is best to spend several minutes in familiarizing yourself with the formation under observation before starting a drawing than to commence drawing immediately.

While you are doing this your eye is becoming adapted to the brightness level of the telescopic image and the contrast range. During prolonged observation the focus of the eye might change slightly and you will have to make slight alterations of the telesocpic focus as a result. The merest turn of the focusing wheel is usually sufficient.

If you have been studying a formation near the terminator and then change to an object near to the limb where the illumination is much brighter, you will probably find that this also will necessitate refocusing the telescope slightly. The same will happen if you switch from the limb to the terminator. The reason appears to be changes in focus of the eye caused by opening and closing of the iris in response to changes in illumination intensity.

When you feel that you have become thoroughly familiar with the detailed appearance of the crater, you may commence drawing. Attach your drawing paper to a clip board and illuminate it with a dim flashlight or better still a dim shielded miniature electric light clipped to the drawing board and connected to the electrical cord that powers the telescope drive. There is no need to use a dim red light as recommended by planetary and 'deep sky' observers. The red light is used because unlike bright white light it does not ruin the 'dark adaptation' of the eye which is so important to observers of faint celestial objects. The Moon is so brilliant that dark adaptation is unimportant when observing it.

Draw lightly so that errors can be corrected easily with a soft eraser. Use a pencil with an eraser attached to the upper end; a loose eraser is easily dropped and you won't want to hunt for it in the dark during a critical observation. Use a medium hard pencil. Depending on how much detail there is to show, a scale of about one inch to 10 miles may be used for the crater diameter. Begin by lightly sketching the outline of the crater as accurately as possible. Time spent on getting the outline right at this stage will save considerable grief later. It is no good filling an inaccurate outline with accurately drawn detail. If the crater has a well-marked bright rim, sketch this next, noting its exact shape and variations in its apparent thickness. Next, draw the outlines of the major shadows, being especially careful to delineate accurately all irregularities in their outlines as these are indicative of the shapes of the structures casting them. Do not waste valuable observing time at the telescope filling in the shadows with black; this can be done later. The area within the shadow outline can be temporarily indicated with say a few small dots, crosses or other symbols, so that when you put the finishing touches later indoors, there cannot possibly be any doubt as to which part of your drawing is meant to be shadow and which light.

Then, indicate the exact position and size of any central mountain or mountains and other objects on the crater floor such as craterlets, hillocks or clefts. Check the accuracy by aligning these features with details on the crater walls in the north–south and east–west directions and also with respect to each other. The proportionate size of floor details must be carefully attended to as it is easy to make mistakes. I personally tend to draw crater floor detail too small and have to be on my guard all the time.

When these details have been drawn, carefully delineate the shadows cast by them. Continue observing the crater even though you think that all detail has been recorded. Occasionally in a moment of extra steady seeing, a hillock or crater you have drawn as a single entity may be seen to be double, or maybe the shape was not quite right. You may also glimpse delicate detail that you had not seen before such as fine cracks or rills. The experienced observer waits patiently for these fleeting moments of superb seeing and makes best use of them. Much nervous energy and time are expended in trying to see and accurately delineate fine lunar detail; patience and perseverance are absolutely essential qualities in the successful lunar observer. F.H. Thornton suggests using sidelong glances through the eyepiece while you are engaged in drawing as this often reveals detail that may be missed if prolonged staring only is used.

After drawing the outline of a fairly large formation, inexperienced observers may tend to concentrate on one small area of the formation in order to record as much detail as possible. The temptation is especially great on nights of 'crystalline' seeing. However, many hours would be required to finish an absolutely complete detailed drawing of even a medium-sized formation such as Hevelius when it is on the terminator. If the seeing deteriorates suddenly you would have to stop work and all you would have is an outline with a tiny bit of detail in one corner. That bit of detail may be quite valuable admittedly but if your aim is to draw the whole formation it is better to 'hop around' the formation putting in the fairly coarse detail at first rather than concentrating on one small area. In this way a more balanced effect is obtained. If you are forced to stop observing for any reason with an incomplete drawing done this way you will at least have a good moderately-detailed outline, which will be aesthetically more pleasing and useful on future occasions when you observe the formation again.

Seeing will usually improve as the Moon climbs in the sky towards the meridian. This will often present opportunities to correct any errors that may have crept into your drawing earlier and as the observation proceeds.

If possible, always sketch in some other named formation near to the object you are drawing so that there can be no possible doubt about the identity of the observed formation.

When you are fairly satisfied that you have recorded all that you can see (your tired eyes or cold hands or feet will help to convince you of this!), quickly check each detail and that nothing has been either grossly misplaced or distorted.

On returning indoors go over all outlines more heavily and fill in all shadows. I have found that the easiest way to do this so that shadows are evenly blackened in is to use a black felt-tipped pen with a circular motion all over the shadow area. I find that attempts to fill in shadows with water color and brush always end up with an ugly blotchy appearance and pencil is not much better. However, less dark areas can best be shaded with pencil and the result is better if the pencilled area is rubbed lightly with the tip of the little finger so as to smooth out irregularities in the shading.

Some people may object to the 'heterogeneous' or incongruous appearance of combined pen and ink shading. This may be eliminated by making Xerox copies of the drawing, the results of which are often more pleasing than the original!

If any feature has been doubtfully glimpsed, then it is best not to insert it in the drawing, or if you do, to indicate it tentatively with dotted lines and make a written note about it. Stages in making a lunar drawing are shown in Fig. 5.1.

When the drawing is complete, descriptive notes should be made about each object recorded so that anyone examining your drawing will be in no doubt as to what you saw. Brief notes may be made directly on the drawing with a thin line or arrow connecting the feature with the relevant note. Code letters or numbers may be used instead if there is insufficient room on the drawing for notes and the notes themselves may be written on a separate sheet against the appropriate code number or letter, or even on the back of the drawing. An example of a complete annotated drawing of the lunar crater Atlas is shown in Fig. 5.2.

Any 'write-up' or notes regarding the formation should be done immediately after the observation. It is surprising how easy it is to forget details even if there is a delay of only a few days between making the drawing and writing it up. Your memory can play tricks and errors are sure to creep in.

The following data should be inserted on every observational drawing; omission of any one of these will detract from the value of the observation.

i *The name of the formation.*
ii *Date and time of beginning and end of the observation.* Express the date

thus: year, month, day. Universal Time (UT) should be used rather than local time. This is reckoned from the Greenwich Meridian (0 degrees longitude). Since the Earth rotates from west to east, the hemisphere of the Earth to the west of the Greenwich Meridian is 'behind' UT and the east half of the Earth is 'ahead' of UT. Since the world is

Fig. 5.1 Stages in the making of a drawing of the lunar crater Schickard. May 27, 1961. F.W. Price. Three-inch refractor.

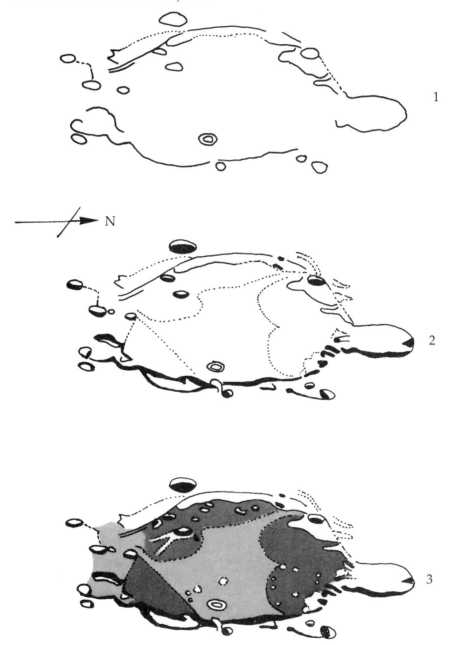

divided into somewhat arbitrary time zones, it would be best to record
your local time and precise location as well as the UT when the observation is made. Supposing that you make an observation from Buffalo,
NY, USA. Buffalo lies in the Eastern Standard Time (EST) Zone of the
USA and its local time is five hours 'behind' Greenwich time. In UT
the day begins at Greenwich at 0.00 hours UT (midnight) and is

Fig. 5.2 Annotated drawing of the lunar crater Atlas.

Formation: *Atlas.* 1974 April 26th. *Colongitude*: 330.0 app.
Telescope: Criterion 8in F/7 reflector
          3X Dakin Barlow with 16mm Brandon ortho. EP. *Power*: 252X
Time: 20.30–21.40 DST. *Conditions*: Sky clear, cool still air
                    Seeing: Quite good, some steady moments initially,
                              deteriorating later.
Observer: F.W. Price.

r^1 and r^2 seen as delicate black
lines in steady moments.
South ends seem slightly
expanded as shown. No other
clefts seen.

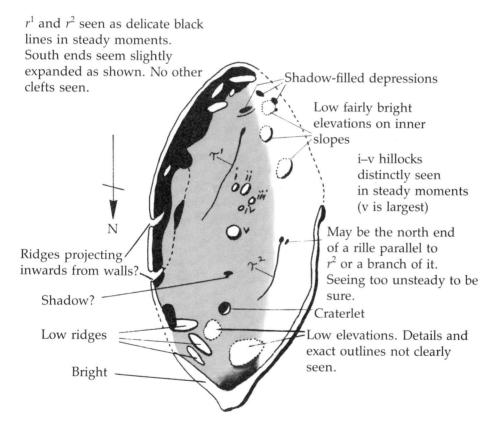

Shadow-filled depressions

Low fairly bright
elevations on inner
slopes

i–v hillocks
distinctly seen
in steady moments
(v is largest)

May be the north end
of a rille parallel to
r^2 or a branch of it.
Seeing too unsteady to be
sure.

Craterlet

Low elevations. Details and
exact outlines not clearly
seen.

N

Ridges projecting
inwards from walls?

Shadow?

Low ridges

Bright

Notes. Observation was discontinued because of deterioration in seeing.
Poor seeing prevented recording of all detail capable of being shown
by the telescope. Fairly certain that no other clefts were visible at
this time. Drawing correlates well with plate B3 in the *Consolidated
Lunar Atlas*.

reckoned by the 24-hour system to the next midnight. Supposing it is 10.30 p.m. EST on February 26 in Buffalo, i.e., 20.30 hours EST on the 24-hour clock. Since Buffalo is five hours behind Greenwich time we must add five hours to this to calculate the UT. 20.30 plus 5 gives 25.30 which is 01.30 hours UT on February 27. (Note, therefore, that the day of the month on which you make an observation may be different when time is expressed in UT.)

iii *Telescope used, aperture, F-ratio, eyepiece type and focal length, whether Barlow lens was used, overall magnification.* For example: '8-in. F/7 reflector, 12 mm orthoscopic eyepiece with 2× achromatic Barlow lens. Power = 224×.'

iv *Temperature and seeing conditions.* Record the temperature in degrees F or C. Mention whether the sky was clear, cloudy or misty; wind, breeze or still air. The 'seeing' is estimated using Antoniadi's scale:

1 *Very good.* Image sharp and quite steady even when high power used.

2 *Good.* Same as (1) but definition occasionally spoilt by slight tremor in the atmosphere.

3 *Fair.* Frequent unsteady periods but good seeing in the steady intervals.

4 *Poor.* Tremulous atmosphere seriously interfering with observation. Steady intervals infrequent and transient.

5 *Very poor.* Unsharp image even with low power and not possible to detect fine detail with any certainty.

v *The position of the Moon's terminator.* In recording the phase of the Moon, i.e., the position of the terminator, it is not enough to state the Moon's 'age' (days after new Moon) or even to quote the phase such as 'full' or 'first quarter'. In the latter case, for example, although the Moon as seen from Earth may be exactly at the first quarter, or dichotomy as it is called, and the terminator is straight, this does not mean that the terminator will be in exactly the same position relative to neighboring formations in successive lunations at dichotomy. This is because of the Moon's longitudinal libration. At mean libration, the great walled plain Ptolemaus will have its west wall just within the light half of the Moon's disc at first quarter whereas at the maximum libration, which exposes the east limb, the entire formation would be visible at first quarter.

A better way is to make an estimate of the position of the terminator relative to a named formation, e.g., 'grazing west wall of Copernicus', or to roughly estimate the longitude of the terminator west or east of the central meridian and to state whether it is the morning terminator (waxing Moon) or the evening terminator (waning Moon).

The best and most accurate way to fix the position of the terminator is to state the Sun's *selenographic colongitude* (usually abbreviated to *colongitude* or *colong.*) This is the longitude of the sunrise terminator measured eastwards from the central meridian all around the Moon. Thus, at first quarter at mean libration the colongitude would be 0 degrees, at full Moon 90 degrees, at last quarter 180 degrees and at new Moon 270 degrees. After the new Moon the colongitude increases from 270 degrees to 360 degrees and we begin again at 0 degrees at first quarter. It follows that the Sun will be just rising on an object whose longitude as measured eastwards from the central meridian is the same as the Sun's selenographic colongitude.

The reason why colongitude is the most accurate way of fixing the position of the Moon's terminator is because it does not have to be estimated; it can be calculated from data regarding the position of the Moon's terminator that have been computed for months into the future and is independent of libration. A colongitude of 0 degrees as seen from the Earth could correspond to a slightly crescentic Moon through dichotomy to a slightly gibbous Moon, depending on the longitudinal libration. The solar illumination angle of any lunar formation is there-fore exactly the same at the same colongitude in any lunation. The precise illumination will be affected also by variations in the libration in latitude. At any given colongitude the telescopic appearance of a formation will therefore be influenced by variations in the longitudinal and latitudinal librations, especially near the limb regions. The colon-gitude at 0.00 hours UT on every day of the year is published in tabular form in the *Handbook of the British Astronomical Association*. For the method of calculating the colongitude at any time on a given day, see Appendix 6 at the end of this book.

vi *The cardinal points.* These should be inserted on the drawing so that the orientation of the formation will be unambiguous. Remember that the north–south and east–west lines will only be perpendicular to each other if the object drawn is at, or very close to, the center of the Moon's disc at mean libration. At all other places, especially near the limb regions, the north–south and east–west lines will be inclined at angles other than a right angle.

One should not be content with a single observation of a given formation; an object cannot be said to have been adequately observed until several drawings have been made of the object under as many different conditions of illumination as possible. In making subsequent drawings there is no need to waste time starting from scratch and drawing the outline of the formation and the coarser details every single time. It is better to prepare a set of outlines upon which shadows and fine details can be added at the telescope. As well as saving time and energy which can be spent on record-ing detail not previously seen, the identical shape and size of the outlines made in this way make comparison of different observations easier than if they had been done freehand each time at the telescope, with the inevitable minor variations in size and shape which would result from freehand draw-ing.

Outlines from your first drawing may be prepared with the aid of a simple copying desk that can be easily improvised. Lay a sheet of thick plate glass about 18 inches by 24 inches or thereabouts across two stacks of books about six inches high. Place your original drawing face up on the upper side of the sheet of glass and lay a similar sheet of clean drawing paper on it with the edges coinciding. Attach the upper sheet to the drawing with paper clips to prevent slipping. Arrange an electric light bulb connected to a source of current beneath the glass plate, switch on and your drawing will be plainly seen through the top clean sheet by transillumination. The outline may then be made by lightly tracing the image of the major details of the drawing onto the clean upper sheet with a pencil. Preparing 'blanks' in this way is acceptable if the formation is not too far from the centre of the Moon's disc, but the shape of formations near the limb are quite markedly affected by libration. Three blanks of such formations should be made so that the shape is about right for the two extremes of libration in longitude as well as mean libration so that on any one night, one or the other of these should be a close enough approximation to the shape of the formation. A good plan is to file away your 'blank' outlines in alphabetical order of the formation names. You will then be able to locate easily any one of the outlines on a given evening without having to waste time hunting for it among the others, as you would if they were filed haphazardly.

Selecting formations for systematic study

There is, of course, enough detail on the lunar surface to keep anyone busy for more than a lifetime even if one owned only a three-inch refractor. You must therefore be selective and choose a few objects to study in detail. A good plan is to select, say, half a dozen or somewhat more formations and to concentrate on these, observing them at every available opportunity, and under as many different lighting conditions as possible. Such a seemingly modest program should provide enough work for several months at least.

The formations chosen should be spread as widely as possible in longitude across the Moon's face from the east to the west limb. If the locations of the chosen formations are thus widely distributed in longitude, then on any one night of the lunation there will certainly be at least one of them visible and almost certainly at least one will be seen under conditions of oblique illumination and maximal contrast. Conversely, if a few objects are chosen within a narrow band of longitude, then on many nights during the lunation, especially if near the west limb, none of them may be visible under conditions of good contrast. If near the east limb then you would have to wait for several days after new Moon before you could even see them and then you may have to stay up late. In both cases many a night of good seeing conditions would be wasted.

As to what to select for study depends on many things including the observer's preferences and interests and his experience. The beginner is advised to stick to formations well on to the disc of the Moon and to eschew the limb regions completely. Formations close to the Moon's limb are seen greatly foreshortened and it is often difficult to interpret exactly what it is that you are looking at. The appearance of such formations is strongly affected by libration and it may be difficult to correlate what you saw on one occasion with what you see on the next, that is if you can even find the object on a subsequent occasion; it may have been carried out of sight beyond the limb by libration. You can safely study formations a little distance from the limb, such as the Mare Crisium, which are not too badly foreshortened. However, they are still strongly affected by libration and such objects should be studied when the libration is favorable, i.e., when the formation is as far as possible on the disc so that foreshortening is minimal. Dates and times during the year when various parts of the limb and near-limb regions are favorably exposed are published in the *Journal of the British Astronomical Association*. Maximal librations in longitude and latitude are listed with dates and times and the limb which is exposed to view.

The ring structures seem to have an especial appeal to the lunar enthusiast and will undoubtedly be chosen by the beginner in preference to any other type of formation for observation and drawing. Practice should first be obtained in observing much-studied craters such as Plato, Gassendi and Aristarchus. Examine published drawings and charts by established observers. When you appear to have acquired skill and reliability in observing and drawing move on to less well known formations. There are plenty of them on the Moon and systematic observations of these could be interesting and perhaps valuable. The sector of the Moon's disc bounded by the south and south-west radii seems to be especially abundant in formations that have not been closely observed or delineated.

Your observations need not, of course, be confined only to craters. There are plenty of other lunar features worthy of study such as rills, curiously shaped hills, domes, the curious dark patches and markings within certain craters and the bright ray systems.

Having finished your drawing and made adequate notes, it is very interesting and instructive to compare your work with the representations of the formation you have just drawn with the standard maps or with individual published drawings. A list of drawings of lunar formations is given in Appendix 8 but it does not pretend to be anything like a complete list. Almost certainly you will notice discrepancies between your drawings and the delineations of others of the same formation. Apart from differences obviously due to the varying angle of illumination in different drawings you may note that you have recorded a structural feature or features in your drawing that is absent from the other. Conversely, you may have missed something that is shown in another's drawing. Again, your delineation of a feature may differ significantly in shape and size from the representation of the same feature in a drawing by some other observer. Relative positions of different objects may differ in drawings of the same formation by different observers. Comparison of your drawings with the work of others is therefore often a revelation and sometimes puzzling but it is these discrepancies between your and another's observations of lunar features that makes lunar observation such an absorbing pursuit. This phenomenon is not, however, peculiar to lunar observation. Telescopic drawings made by reputable observers of the same planet at the same time and using the same aperture and magnification often show marked differences that are difficult to explain. These and lunar observational discrepancies can, in the vast majority of cases, be ascribed to personal error and to differences in seeing conditions and librations. Perhaps also they arise from physiological or psychological reasons in some cases. Some people have sharper vision than others and interpretations of what is seen may differ from one observer to another. When you encounter observational discrepancies, as you almost certainly will, do not immediately conclude that you have witnessed a cataclysmic change in the structure of a lunar formation or that you have made a momentous discovery! From time to time you may find that something you have observed or drawn on the Moon looks different from how you observed or drew it weeks or months previously. Herein lies the importance of scrupulous objectivity and honesty in recording and drawing. If you do notice something strange or different about the appearance of an object, compared with what you represented in a previous drawing, that is not attributable to seeing conditions, libration or angle of illumination, then you can be more confident in the belief that there may really be a difference; you know that you can rely on your previous drawing if you have consciously disciplined yourself to draw only what you see and to firmly avoid putting in any feature that is doubtful, especially if it is at the limit of resolution of your telescope.

As to what the difference is due to is another matter; it may after all be only that your skill in delineation and observing ability has increased since your first drawing. It is doubtful whether sructural changes occur today on the Moon that are discernible in even large telescopes.

Your finished drawing as made at the telescope may not be a finished artistic masterpiece but it is the direct result of observation; as such, it is valuable and should be stored in a file of your original drawings. Absolutely nothing should be added, subtracted or altered on it once you have left the telescope, except to shade in shadows or to go over outlines more heavily. Although the temptation is sometimes great to 'touch up' something to 'make it look better', or to add detail that should be there and was not seen, it should be steadfastly resisted.

As well as your original, you may also wish to keep a book of accurate copies of your original made on the copying desk that have been properly finished, outlined in a frame and devoid of labels and arrows, which are more suitable for publication if you aspire to this, for display purposes or for showing to friends. These need only have the name of the formation, date, telescope used and observer's name neatly stencilled at the top or bottom of the drawing. A drawing for publication should have all relevant data, of course.

Using the telescope efficiently

When viewed under excellent seeing conditions, large formations such as the walled plains Clavius or Grimaldi are seen to contain an enormous amount of delicate detail when viewed with an eight-inch reflector. To attempt drawing from scratch such formations showing every detail at one sitting is virtually hopeless. They are quite a challenge with even a three-inch refractor. By the time you have drawn just the major features and got the proportions and shapes correct you will be probably too tired to go struggling after the delicate details. However, such an incomplete drawing should not be cause for despair; it still can be used as the basis for inserting the fine details when the formation is observed on another occasion. A better plan is to find a good photograph of the formation and to trace or accurately copy an outline of the formation in as much detail as the photograph shows, but not including the shadows. When you next view the formation check all the details of the drawing at the eyepiece with what you can actually see and satisfy yourself that you have confirmed every detail in the drawing. Next, insert shadows as they actually are on the occasion of the observation and erase detail in the drawing if hidden by shadow. Then, go after the fine detail; insert everything that you can certainly see on the drawing.

Some purists may look upon sketching a detailed outline from a photograph as 'cheating' and say that all lunar drawings should be from scratch. I used to think this at one time but there is no need to feel this way. If you have honestly seen and confirmed at the telescope all the details in your prepared outline and inserted corrections where needed, then your outline is an honest observational drawing. All that the photograph has done is to save you much time and patience in getting major proportions and dimensions correct. Your main energies may then be devoted to delineating the finest detail that your telescope can show. After all, that is presumably why you are using an eight-inch rather than a three-inch telescope – to see detail of a certain degree of fineness. If you use an eight-inch telescope for habitually sketching detail that can be easily seen in a three-inch then the eight-inch is not being efficiently used.

On nights of mediocre seeing, the eight-inch cannot be used efficiently in this sense because fine detail may not be visible anyway. However, there is no need to dismay. You can still use the opportunity to draw the outline and coarse details of a formation that you have not previously observed and use a later occasion under better seeing conditions to make the drawing more complete and detailed; it would be a waste of observing time to use a night of really fine seeing merely to make an outline.

Thus, the telescope may be used at various degrees of efficiency determined by the seeing conditions of each observing occasion and by so doing you get the most out of your telescope and the time spent in using it. It is fortunately rare that seeing is so bad that no observation of any sort is possible.

The owner of an eight-inch telescope might well concentrate on detailed studies of smallish craters like Timocharis, Lambert, Pytheas and Proclus. These are not seen to advantage with a three-inch refractor but with an eight-inch they really 'come to life'. On nights of excellent seeing your eight-inch reflector will be efficiently used if you decide upon systematic observation of craters in this size range.

As previously mentioned, the best and most steady seeing often occurs on clear cold winter evenings. At this time of year the Moon is quite high in the sky when on the meridian and this also is conducive to good seeing. Such nights obviously present excellent opportunities for observing and recording much fine detail on the lunar surface. Unless you are warmly clothed and your feet well insulated, you will begin to feel stinging cold in your toes and ankles after about 20 minutes or so of observing. If you are drawing a formation your fingers will quickly suffer too. It is better at such times not to attempt anything like a complete drawing of a fairly detailed formation that may require an hour or more of hard work. The time would be better spent in short periods at the eyepiece checking fine or doubtful detail on one or other of your more or less complete drawings of a lunar feature that is well placed for viewing. There is always the chance that what has always looked like a hill or craterlet may resolve into two very close small hillocks or craterlets; or you may be able to trace a crack or cleft further than you hitherto have been able. You may even perceive detail that you have not seen before. If you do decide to make a drawing on a freezing night, select a formation about 20 miles in diameter or thereabouts such as the craters Timocharis or Lambert. Under fine seeing conditions a good delineation of features of this size need not take more than half an hour. Perhaps a good principle might be to proceed with 'more haste and less speed' on such cold nights.

Highlight observations

Most textbooks and articles on observational astronomy warn that the worst time to observe the Moon is when it is full, owing to the lack of shadows and contrast, and the glaring brilliance of everything. Many large formations are indeed practically invisible at full Moon and their dramatic and rugged appearance when on the terminator is utterly gone. However, complete observation of a given formation must include its appearance under highlight conditions as well as the more exciting and dramatic views when the Sun is at a much lower angle.

Observing lunar features under highlight conditions is admittedly rather trying and takes no little self-discipline when we know that a touch of the telescope adjustments will bring into view much more exciting and detailed Moonscapes near the terminator where the glare is less trying on the eyes. The diminished contrast is worsened by the central obstruction of a Newtonian reflector although I have found that this effect is not as bad as some writers seem to imply. However, the effect of slight atmospheric tremor, which may be intolerable when viewing features under the high contrast conditions of low angle illumination, does not seem so bad when studying features under highlight conditions. Nevertheless, the close study of lunar features at full Moon is likely to be more physically trying and mentally exhausting than when they are on the terminator.

A good way of getting the 'feel' of highlight observation is to study the Mare Crisium when the libration is favorable. This will be illuminated by a high Sun from somewhat before first quarter to full Moon and so is 'available' for highlight study for several days without you having to stay up too late. This isolated mare seems to have a special fascination for English lunar observers. Its surface is covered with spots and streaks and the fan-like rays from the nearby crater Proclus.

As a start, the group of white spots knows as the 'Trapezium' or 'Barker's Quadrangle' in the southern part of the mare near the Cape Agarum should be studied. This 'formation' is well seen even with a three-inch refractor at a power of 90× on a still clear evening. Curiously, the Trapezium, which is glaringly obvious in such a small telescope, is omitted from some of the earlier lunar maps such as those of Beer and Mädler.

Sketch in the outline of the Cape Agarum as a reference feature and then carefully study the number and configuration of the white spots and any streaks that you see. Plot each spot accurately with respect to the outline of the Cape and check all alignments, both of the spots with themselves and in relation to the Cape Agarum. You may be surprised at how easy it is to make positional errors in drawing such a seemingly simple object. Then compare your drawing with Figs. 5.3 and 5.4. Fig. 5.3 is the first drawing I made of the Trapezium with the aid of a three-inch refractor at 90×. Several years later I drew it again, this time observing with an eight-inch refractor at 250× (Fig. 5.4). In the meantime I had forgotten the configuration of the spots so my later drawing was unbiased by previous knowledge. Notice the excellent agreement and the comparatively little further detail revealed by the larger aperture. My drawings agree completely with the best available photographs, yet, strangely, drawings by other observers do not entirely agree with my own observations.

Having satisfied yourself that you have observed the Trapezium accurately, try constructing a chart of the entire Mare Crisium using a prepared outline. A good size would be about 10 inches for the north–south extent of the mare. The chart of the highlight features of Mare Crisium shown in Fig. 4-44 was constructed from about three or four partial charts made on different evenings. It does not pretend to be exhaustive. Observe carefully the white spot west of the crater Picard and the rays from Proclus, as they have been said to be 'variable'.

Fig. 5.3 The 'Trapezium' in Mare Crisium. November 23, 1961. F.W. Price. Three-inch refractor.

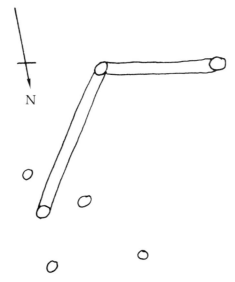

Under conditions of high illumination, most or all of the rugged details both coarse and fine that were visible in, say, a crater when near the terminator will disappear under high light, but new detail of a different sort will have made its appearance. A previously uniform-hued crater floor may now be variegated with light and dark spots, patches or streaks. These will usually begin to appear soon after local sunrise and be fully 'developed' under high light.

Desultory versus systematic observation

To make the best use of observing time you should go to the telescope with a definite idea of what you want to accomplish. Desultory viewing rarely, if ever, yields any results of value. On a clear cold evening the desultory observer may look out of the sitting room window on a clear winter evening and note that the Moon is bright and high in the sky, so out comes the telescope. After a few minutes of wandering up and down the terminator

Fig. 5.4 The 'Trapezium' in Mare Crisium. February 22, 1964. F.W. Price. Eight-inch refractor

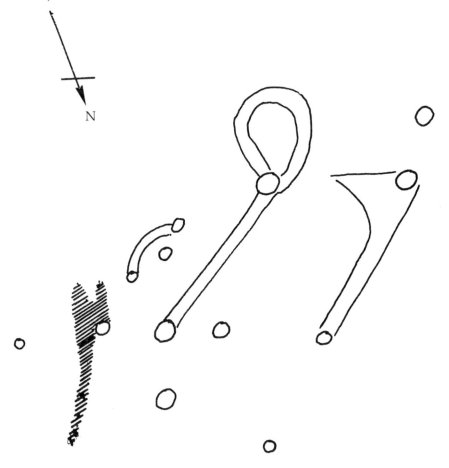

N

he spots something that takes his fancy and tries to decide whether it would be worth making a drawing. The seeing is perhaps tremulous (as it usually is) and fine detail is difficult to 'hold'. Finally, as the toes and fingers begin to get cold a drawing is decided upon. After all this indecision, the observer is determined to 'get something down on paper'. After a few minutes the drawing does not seem to be going well and the by now irritable observer decides to stop, as with freezing fingers he screws up the unfinished drawing in disgust and tosses it into the waste paper basket on getting indoors.

A far better plan is to know in advance the position of the Moon's terminator on a particular evening by calculating the colongitude. Then consult a map of the Moon and decide which formation or formations will be favorably placed for observation and decide exactly what it is that you intend doing. This may be to check detail in a previous observation or to make a new drawing. Planned observation is much better than going to the telescope with little idea of what you are going to do.

Care of eyepieces

Good eyepieces are expensive and should be handled carefully. When not in use always keep them in the containers in which they were sold. For more convenient and rapid access you may want to keep them all in one box. This should be fitted with a dust-proof lid. To prevent the eyepieces from moving around inside the box the latter should be fitted with a shelf in which circular holes have been cut to hold them. The holes may be lined with velvet or other soft material. Another plan is to fit the box with a thick pad of some kind of foam rubber or other plastic with circular holes cut in it for the eyepieces.

During observation never carry extra eyepieces unprotected in your coat pocket but in their individual carrying cases or in small plastic bags.

Never touch the lens surfaces with the fingers or attempt to clean them by wiping with any cloth that comes to hand. Always use a soft camel hair brush to remove dust. If further cleaning is needed use only lens tissue or a very soft well-washed handkerchief that is reserved for this purpose only.

If your eyepiece lenses are coated, sooner or later the bluish sheen on the lens surfaces may become 'moth eaten' in appearance. This is no cause for alarm and does not affect the performance of the eyepiece.

Making charts and maps

A drawing of a lunar formation made at one particular time represents that formation as it appears at one particular angle of illumination. Details that are plainly visible under one condition of illumination may be invisible under another, apart from the obvious effect of shadows in hiding objects. The lunar observer soon realizes how the appearance of a given formation is strongly dependent on the illumination angle. A whole series of drawings would be needed to portray adequately any lunar formation.

The details revealed in such a series of drawings can be combined in a single drawing so as to make a map or chart of the formation. If such a chart is to be reliable the formation should be critically examined from sunrise to sunset over several lunations. Much patience and time are therefore required for this kind of work. Fine details should be searched for as soon as the Sun rises on them because quite often they disappear as soon as the Sun climbs a little higher. Fleeting detail at sunset should be similarly watched for.

Needless to say, shadows are not put on a map or chart. Some formations are more convenient to observe than others. Formations in the eastern half of the Moon's disc will not be seen under sunset conditions until perhaps well past last quarter. This will mean staying up into the early hours of the morning and then observing the Moon fairly low down in the sky. Formations in the western half of the Moon's disc are not so trying in this respect. If in longitude about 30–40 degrees west, they can be well observed at lunar sunrise as the crescent Moon at this age (about four days after new) will be quite high in a dark sky after earthly sunset. The same formations will be observable under lunar sunset conditions three to five days after full Moon, which shouldn't keep you out too late at night especially in the winter, when the clocks have been 'put back' after being 'faster' than mean time by one hour during the spring and summer. Such formations as Frascastorius, Capella, Isidorus, Piccolimini, Atlas, Hercules and Posidonius fall into this category.

Further Reading

Articles

'The Art of Lunar Drawing.' Chapman, C.R. *The Strolling Astronomer,* **17** 45–9 (1963).

'Notes on Drawing.' Thornton, F.H. *The Moon* (BAA Lunar Section Bulletin), **3** (2), 43–4 (1954).

'A Note on the Accuracy of Drawings and Maps.' Brinton, H. *The Moon* (BAA Lunar Section Bulletin) **9** (1), 6–8 (1960).

'Notes for Novices', no. 1. Thornton, F.H. *The Moon* (BAA Lunar Section Bulletin), **7** (4), 74–80 (1959).

'Notes for Novices', no. 2. Rackham, T.W. *The Moon,* (BAA Lunar Section Bulletin), **8** (2), 27–9 (1960).

'Notes for Novices', no. 3. Rae, W.L. *The Moon* (BAA Lunar Section Bulletin), **8** (4), 75–7 (1960).

Volumes of the British Astronomical Association Lunar Section Bulletins *The Moon* may be found in the Association's Library at Burlington House, Piccadilly, London W1V 0NL.

6

Mysterious Happenings on the Moon

The most casual telescopic observation of the Moon shows that the appearance of the surface features changes from night to night, even from hour to hour. The continually changing angle of solar illumination causes striking cyclic alterations in the visibility and appearance of lunar formations but these are not, of course, real physical changes but are due only to the constantly changing interplay of light and shadow. They are accurately predictable.

From time to time, ever since serious scientific observation of the Moon began, there have been reports of localised changes and other mysterious happenings on the Moon's surface. They are usually transient and cannot be related to or explained in terms of solar illumination angle or libration and are unpredictable. However, there has never been anything as catastrophic as, say, the appearance or disappearance of a mountain range or a walled plain. In fact, there is only one case of an apparent permanent physical change, and this is relatively minor, plus a few other doubtful ones.

The vast majority of well-documented 'happenings' are temporary alterations in the appearance or visibility of localised areas or formations, and transient flashes, lights or glows that last for a second or so or for several minutes.

The 'changes' and mysterious happenings on record that deserve serious consideration are very numerous and cannot all be described here, but a few examples will give the general idea. Unusual happenings on the Moon fall mainly into four categories:

    i apparent physical changes – mostly disappearances and appearances of small craters;
   ii clouds, mists and obscurations;
  iii variability in light and dark spots and streaks;
  iv lights, flashes and glows.

Disappearances and appearances

Linné. The most celebrated case of an apparent physical change on the Moon's surface is the alleged disappearance of the crater, Linné. The announcement of this in 1866, by J. Schmidt, caused quite a stir in the astronomical community because for many years it had been believed that the Moon was a dead and unchanging world and lunar studies were therefore at a standstill. Paradoxically, this was the result of a major advance in the field of lunar science, the publication in 1837 of W. Beer and J.H. Mädler's *Mappa Selenographica* (Selenographic Map, Map of the Moon), and the accompanying book *Der Mond* (The Moon). These were the results of years of painstaking telescopic scrutiny of the Moon and accurate mapping of the surface features. Mädler did most of the observing using the 3¾-inch Fraunhofer refractor in Beer's private observatory in Berlin; Beer's contribution was quite considerable too, and must not be ignored. Mädler's reputation was so much respected in astronomical circles that the publication of the lunar map and book, which were the most detailed and accurate at that time, and Mädler's declaration that the Moon was a dead and changeless world were considered the final word. This had a deadening effect on selenography because it seemed unlikely that anything new remained to be discovered about the Moon.

Much has been written on the subject of the disappearance of Linné, but this chapter is not the place to review the voluminous literature on what has become known as the Linné affair. However, the main facts are as follows:

In moderate telescopes, Linné appears today as a small whitish spot upon which is a tiny craterlet or pit about one quarter of a mile in diameter (fig. 6.1). It is situated in an isolated position on the Mare Serenitatis at approximately latitude 27° 47' north, and longitude 11° 32' west.

According to Pickering, the earliest known record of the appearance of Linné is on Riccioli's map of the Moon (1651), in which it is shown as a moderately-sized deep crater. In 1788, Schröter described it as 'a very small, round brilliant spot, containing a somewhat uncertain depression'. In 1823, W.G. Lohrmann described Linné as a crater about 4½ statute miles in diameter and the second most prominent crater on the Mare Serenitatis (the most prominent is Bessel which is 12 miles in diameter and more than 3000 feet deep). J.H. Mädler observed it seven times and represented it as having a diameter of seven statute miles. Both Mädler and Lohrmann measured and drew it as a point of reference when constructing their lunar maps. J. Schmidt made five drawings of Linné between 1841 and 1843 showing it as a crater; apparently he was the last to see Linné with these characteristics for on the evening of October 16, 1866, he was observing the Mare Serenitatis when to his astonishment he realised that Linné had disappeared. All that was left of it was a small whitish spot.

Schmidt's announcement of the disappearance of Linné caused a worldwide sensation in astronomical circles. Many were the telescopes pointed at Linné and many were the drawings made of it during the next few years.

Although they were not all in complete agreement, it was apparent that Linné was nothing more than a small white spot upon which was a tiny object variously described as a hill or a craterlet. Subsequently, Secchi and others saw a shallow depression six miles across within the white spot of Linné that contained a tiny pit. Corder, and later Goodacre, represented Linné as a crater cone upon the eastern rim of a shallow ring, on the western rim of which was a small peak. The shallow depression now seems to have disappeared and in its place is now a craterlet surrounded by a white patch.

Pickering decided that the original Linné had disappeared between 1843 and 1866, and the shallow depression that was seen in its place during 1867 and 1868 had disappeared by 1897. Many other curious 'changes' and 'variations' in its aspect have been detected over the years.

Credibility for the story of the disappearance of Linné depends completely on the accuracy and reliability of the statements made by Lohrmann and Mädler in earlier times and the later ones of Goodacre and Corder. A study of Lohrmann's records suggests that he observed Linné only under a high Sun and incorrectly interpreted the white spot as being the bottom of a crater. It appears that Mädler made the same mistake; Linné was only one of several white spots that he used as primary reference points in fixing the positions of lunar surface features and many, but not all, of these white

Fig. 6.1 Linné (based on Fauth's chart). The dotted ellipses represent the white spot on which the Linné craterlet stands. F.W. Price

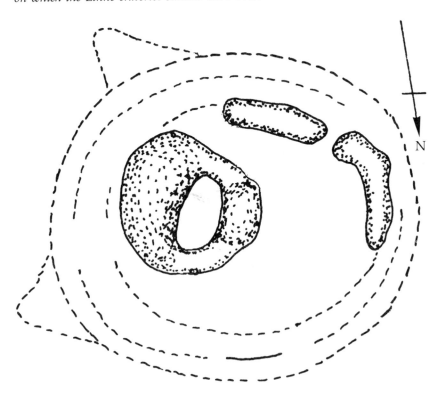

spots are really craterlets and are only seen as such under low lighting. It should be mentioned here that the seven times that Mädler observed Linné does not mean that he attached any great significance to it. The observations were all made on the same evening under a high Sun and were all determinations of the position of Linné. So Mädler only actually saw it under one set of illumination conditions, in which it appeared as a white spot. Therefore it seems probable that neither Lohrmann nor Mädler actually saw Linné as a crater but assumed it to be and recorded it as such. In fact, Mädler incorrectly recorded two other white spots as craters – Lassel D (the former Alpetragius d) and Birt C.

Further doubt is cast on the alleged change in Linné by differences in the two published accounts of Schmidt's observations. In one, he says that he had drawn the Linné area eleven times in 1841–3 and in only five of these (previously mentioned) Linné is shown and always as a crater. However, in his book published in 1878, nineteen drawings are listed between 1841 and 1843 and in two of these, at least, Linné is shown as a light spot. We must also not forget that when Schmidt made these observations he was quite young (born 1825) and had not the necessary experience or skill that he later acquired; also, he used small telescopes. Therefore, not much reliance can be placed on the observations he made in 1841–3 because of this and the discrepancies of his two accounts of them.

Additional doubt is cast upon the 'case' for a change in Linné by many earlier observations of its appearance as a white spot. Schröter's 1788 drawing of the Mare Serenitatis shows a white patch in about the right position, and the accompanying description supports its identification as Linné. In 1904, A.A. Rambaut drew attention to two of the English artist John Russel's drawings of the Mare Serenitatis, also made in 1788, in which Linné is shown as a white spot.

Finally, Mädler himself declared in 1867, the year after Schmidt's announcement of the 'disappearance' that Linné looked the same to him as it did 37 years previously. Taking all this evidence into consideration, it seems doubtful that a real physical change ever occurred in Linné. Today, Linné appears as a low mound on the summit of which is a deep pit. This is the description given by Thornton, who has examined Linné with his 18-inch reflector. This appearance was confirmed by Wilkins and Moore in 1953 with the 33-inch Meudon refractor (Fig. 4.12). As for the apparent changes in aspect observed by Goodacre and Corder, it can only be said that Linné certainly undergoes some curious apparent variations with the changing angle of illumination.

It may not be merely coincidental that Schmidt announced the disappearance of Linné at about the same time as he was contemplating publication of his lunar map; the renewed interest in the Moon that was the result of the Linné story no doubt would have had a beneficial effect on sales of the map which, however, because of delays, did not finally appear until 1878. Whether or not one believes that Linné disappeared, Schmidt's announcement, in 1866, of the disappearance revived interest in the Moon and lunar studies have been pursued vigorously ever since.

Alhazen Another example of an apparent disappearance refers to a formation on the west border of the Mare Crisium that was described by Schröter as 'a large distinct crater with bright walls and a dusky floor, visible under all lighting conditions'. He measured the diameter as 23 miles and named the crater Alhazen. Most accounts then go on to say that Beer and Mädler could find no trace of it. In its place was an ill-defined depression situated between two mountain peaks. Beer and Mädler therefore transferred the name Alhazen to another crater some distance to the south so that the present day Alhazen is not the same as Schröter's.

A clue to the apparent 'disappearance' is given in W.R. Birt's detailed account of Alhazen in the *Astronomical Register* (Vol. 5, p. 170). He relates how Kunowsky could not find Schröter's Alhazen in 1825, but that it was seen by Pastoroff and Harding in 1827. Then later again, as we have seen, Beer and Mädler could not find it. Birt himself says that in 1862, he found it to consist of two mountain ranges that sometimes looked like a crater. Then, in an observation made with a 4½-inch refractor, on July 5, 1867, under favorable conditions, he ascertained that Schröter's Alhazen is really 'a crater situated on the surface between two ranges of mountains and but slightly depressed below it'. It appears, then, that only under certain critical conditions of illumination and libration does the object have the appearance of a crater. Its crateroid appearance is probably rare and this could account for its apparent 'disappearances' and 'reappearances'. It is situated close to the limb and its form is bound to be greatly affected by libration and illumination angle. Birt says that sometimes it looks like a depressed grey surface within a ring, and at others it looks like a longish flat ridge. Most curiously, remarks Birt, is its 'general *indistinctness* (Birt's italics) while neighbouring objects have been well defined'. Then again, Birt found it quite distinct on some occasions when near the limb. Although it does not seem that Schröter's Alhazen really disappeared, there is, nevertheless, something odd about this object and its variable appearances.

Cassini. Although not seriously cited as an example of a 'new' crater on the Moon's surface, it is yet strange that the crater Cassini should have been omitted from the older lunar maps. It was first represented in J.D. Cassini's map of 1692, and was named after him. The crater Cassini is a conspicuous object under low lighting, but it is by no means prominent at other times; this is probably why it was left out of the older maps. It is definitely not a 'new' crater.

Cleomedes. On the floor of the walled plain Cleomedes is a small crater that Schröter considers came into existence around October, 1789. He arrived at this conclusion because he had not seen it before, and now it was quite an easy object. In fact, he thought that the crater was formed under his eyes while observing Cleomedes. He says that he saw something 'eddying' on the interior and then there it was – the new crater. Schröter thought that he had witnessed a real volcanic outburst. The craterlet is not very prominent and Schröter may have overlooked it on previous occasions, or it may have been obscured by something; mists have been reported in this area from time to time. The nature of lunar 'obscurations' will be discussed later.

Craterlet in Halley. Obscuration may have played a role in what seems to be the appearance of a 'new' craterlet on the floor of the crater Halley since Lewis Rutherford photographed it in 1865. Lohrmann and Mädler showed no floor detail in Halley, but in Rutherford's photograph one craterlet is distinctly shown. This in itself is not evidence of a new craterlet being formed, but what is interesting is that today there are two craterlets on the floor of Halley. It is the second craterlet not shown in Rutherford's photograph that is thought to be new. However, it is possible that it may have been obscured by something when the photograph was taken.

Hyginus N. In 1877, Klein observed an object on the Moon that he thought was new as he had never seen it before. He had studied the region of the Hyginus cleft on many occasions during the previous 12 years, and on May 27, 1877, he saw what he described as a rimless depression close to the Hyginus cleft about three miles in diameter. Under low angle illumination it was filled with shadow. The object was not shown on any of the available lunar maps. The region had been drawn more than 30 times by Schmidt during the previous 30 years and in the position of Klein's depression – which became known as Hyginus N – he sometimes showed nothing, and at others a small bright or dark spot. Klein was therefore certain that a change had occurred, and that he had discovered a new formation.

Not far from Hyginus N is a smaller depression also thought to have been recently formed. Mists have been seen in the area at times and it is possible that the two objects may have been obscured when Mädler and Lohrmann mapped the region. Evidence of localised 'activity' in this area is provided by H.P. Wilkins' observation of the unusual dark tint of Hyginus N on April 4, 1944, and the bordering of the south edge of the Hyginus cleft by a thin dark band for a distance of more than eight miles. In the *English Mechanic* (Vol. 28, p. 369, 1879), there is a set of drawings by Lord Lindsey of the area near the crater Hyginus which show markings that altered their form in twenty minutes. Such changes would seem to be too rapid to be explainable on the basis of changing solar altitude. In the same volume of the *English Mechanic*, on p. 562, Birt says that there were strong suggestions of rapidly occurring changes during a single evening in the region around Hyginus and that the black tint in some areas increased appreciably in 30 minutes; Hyginus N disappeared between 6.45 and 8.00 p.m.

It seems that the evidence for Hyginus N being 'new' is not too convincing although it is certain that curious appearances and happenings are occurring in this region, which is well worth studying (see Fig. 6.2).

In this and other cases of supposedly 'new' formations, it is important to realise that merely because Mädler, Schmidt, Lohrmann and other skilled observers of the past failed to record various minor lunar surface features that are visible today is not evidence that these features have come into existence since their time. Mädler, for example, had a well-deserved reputation for skillful and meticulous observation, but authors who make statements such as 'if such-and-such a feature existed in Mädler's time, surely he could not possibly have failed to see it and record it', are forgetful of two things. First, Mädler was engaged in the colossal task of mapping the whole lunar surface within a few years. The omission of comparatively minor detail here and there should therefore not occasion surprise or constitute grounds for speculation about the appearance of features since his time. Second, he set a lower limit to the resolution of his map so as not to overcrowd it with detail. Therefore, small objects visible in his telescope but below this limit would be omitted. He may even have left out minor features in areas crowded with fine detail so as to maintain a uniform density of detail over the entire map.

Fig. 6.2 Hyginus N (arrowed) and the surrounding region. Redrawn after H.P. Wilkins. 33-inch refractor (Meudon) (from Our Moon, *Muller, 1954)*

H=Hyginus, S=Schneckenberg Mountain

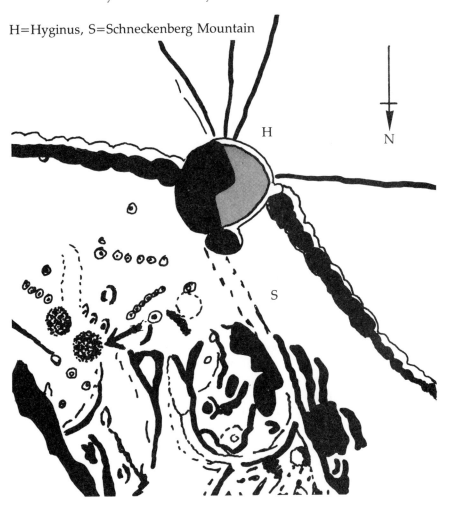

Mädler's square. In the *Mappa Selenographica*, Beer and Mädler, and later Neison, in his book *The Moon*, depicted a remarkable prominent and geometrically regular square-shaped enclosure foreshortened into a lozenge shape immediately west of the crater Fontenelle north-east of Plato on the north 'shore' of the Mare Frigoris (Fig. 6.3). Beer and Mädler themselves remark that the formation and its rampart-like boundaries 'throw the observer into the highest astonishment'. The appearance is so regular as to look artificial until we recall that the walls are 64 miles long, 250 to 3200 feet high, and a mile or so thick. Mädler's Square, as this enclosure has been called, is a striking example of a general parallelism in the surface features between Fontenelle and Timaeus.

Neison describes the square as being enclosed by long straight walls and specifically says that the south-east wall is uniform, straight, and 200 feet high. On the interior, two cross-shaped objects are mentioned, only one of them being mentioned by Mädler. In 1950, Dr J.J. Bartlett drew attention to the fact that the appearance of Mädler's Square does not now correspond to the descriptions of Neison, Beer and Mädler. There is now no clear-cut geometrical appearance and the south-east wall no longer exists. In its place is a low ridge. The present area is no more than a rough part of the lunar surface incompletely delimited by ridges. H.P. Wilkins observing with the 33-inch Meudon refractor also found that one side of the square was missing and the crosses seemed to have disappeared, but they may have been hidden by a shadow at the time.

These observations have generated much discussion and controversy about whether a real physical change has occurred. P.A. Moore, who suggested that the square be re-named after Dr Bartlett, examined the problem. He found a photograph taken by Draper, in 1863, and one of

Fig. 6.3 Mädler's Square (Bartlett) and vicinity as shown in Neison's map

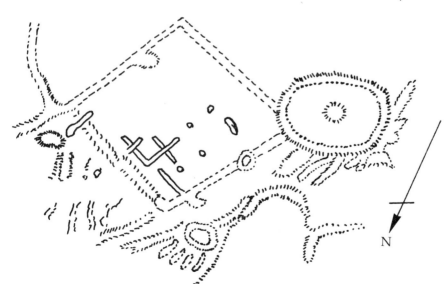

N

215

Schröter's drawings, dated 1809, both of which show the square with its present-day appearance. Neison's book was published in 1876, yet it shows and describes the square as still remarkably regular and artificial-looking. It would seem that Neison never actually observed the area accurately and used Mädler's account for his own amplified description. It appears then that Neison copied an error made by Mädler. P.A. Moore showed that the error could have arisen because of the rather small telescope used by Mädler, a 3¾-inch refractor. He found that when the square is observed with a three-inch refractor it does look like a square, probably due to its 'floor' being somewhat darker than its surroundings, i.e., it is a contrast effect. The same effect could have misled Mädler into incorrectly interpreting the low ridge on the south-east as a wall. The surface to the west is darker than it is to the east and so gives the impression of the square being complete. However, in a 12½-inch reflector, says Moore, the impression of a square is no longer given and the true appearance is revealed. Schröter used a larger though poorer telescope than Mädler and this is probably why his drawing shows the correct appearance of the square. The crosses on the interior have been seen again from time to time. Webb was unable to find the main one ('Mädler's Cross'), but mentions that the Irish astronomer, Birmingham, was more successful. Some American observers, including Dr. Bartlett himself, have seen hints of them. The explanation for the appearances and disappearances of the crosses is probably the same as that for the crater Alhazen. Mädler's Square and Schröter's Alhazen are both situated quite near the Moon's limb, and hence their aspects are strongly affected by libration as well as illumination angle. As previously stated, a critical combination of illumination angle and libration state is required for Schröter's Alhazen to appear like a crater. Likewise, it is quite possible that the crosses in Mädler's Square are visible only when the illumination angle and libration are just right. As shown in Chapter 2, quite a long time elapses between occasions when an object on the Moon's surface can be seen under strictly identical conditions of both illumination angle and libration. In fact, if an observation of a feature is made under a given set of illumination and libration conditions, it may be that poor seeing conditions or clouds will prevent observation of the feature when it is next visible under exactly the same conditions. It is quite possible, therefore, that any one observer may see a particular lunar formation under given illumination and libration conditions only once in a lifetime. Hence, if certain objects such as the crosses in Mädler's Square can only be seen in a very narrow range of illumination and libration conditions, they may well seem to have disappeared to an observer who was fortunate enough to first see them under just the right conditions and then fails to see them on subsequent random occasions. It seems almost certain, then, that there have been no physical changes in Mädler's Square. Nevertheless, the area is worth careful study.

Messier and Pickering. Although the following is not an account of the appearance or disappearance of a lunar formation, it is included here because the phenomena are of a topographical nature.

Messier and Pickering (formerly Messier A) are two small craters situated close together in the Mare Foecunditatis. Pickering is the head of the 'comet tail' formed by the two white streaks extending from it in an easterly direction (Fig. 6.4). The Messier–Pickering crater pair was referred to as the 'odd couple' in a previous chapter because of the strange apparent changes of size and shape that they exhibit at different times.

Beer and Mädler observed and drew the area more than 300 times between 1829 and 1837 and, according to them, Messier and Pickering are exactly alike in size and shape. Today, this description is no longer applicable; Messier and Pickering are quite dissimilar. This has frequently been cited as evidence of change, but the 'case' for change is weak owing to the pronounced variations in aspect undergone by the two craters during a lunation. The variations are due at least in part to the changing angle of illumination.

P.A. Moore made over 500 drawings of Messier and Pickering and, like Beer and Mädler, often found the two to be equal in size and appearing as white spots under a high light. Sometimes, however, Pickering looked smaller than Messier although, generally, it seemed larger. Moore says that Pickering has a triangular shape and is deeper and better defined whereas Messier is elliptical. M. Guest, on the other hand, finds the reverse – Messier is triangular and Pickering elliptical and at these times Messier looks smaller than Pickering.

In 1932, W. Goodacre measured them and found a diameter of eight miles for Messier and seven miles for Pickering even though Messier frequently appears to be the smaller of the two.

A possible cause of these puzzling variations is that Pickering has a curious ear-like projection or extension on the east side. When this is filled with shadow, the rest of Pickering looks elliptical, but with the major axis

Fig. 6.4 Messier and Pickering. November 12, 1954. Eight-inch reflector. Redrawn after K.W. Abineri (BAA Moon, 3(3), 54 (1955))

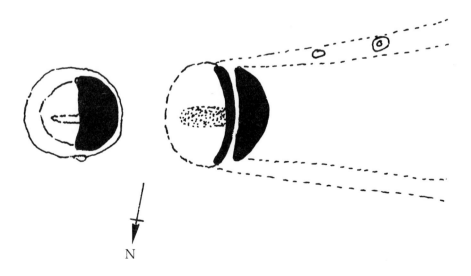

N

in the north–south direction. At other times, when the entire crater is illuminated, it appears larger and with the major axis in the east–west direction. P. Fauth drew attention to the fact that before Beer and Mädler's delineations of Messier and Pickering, F. Gruithuisen had given a more accurate representation of the shapes of the two craters in 1824, and that both Lohrmann and Schmidt represented them with characteristic shapes. Yet, in 1884, Klein showed them in the inaccurate older version of Beer and Mädler.

Lighting effects may not be solely the cause of the variations in Messier and Pickering during a lunation. Some kind of 'activity' is suspected to be associated with them. Klein saw Messier filled with mist on two occasions which seemed to come up from the floor and overflow onto the west wall. P.A. Moore saw the two craters 'strangely blurred' several times and he saw a bright white patch within Pickering on August 20, 1951, 'so prominent that it could not possibly be overlooked'.

It is interesting to note that Moore found that the close pair of small craters Beer and Feuillée in the Mare Imbrium also exhibit apparent variations in size during the lunation. Whatever is the cause of the apparent or actual changes in the aspect of Messier and Pickering, they certainly are worth careful study.

Domes. As mentioned earlier, just because certain kinds of lunar surface features or individual formations were omitted from early maps of the Moon or drawings is not evidence that those objects, now visible, have come into existence since the old records were made. However, it is strange that certain classes of objects that are easily visible today in moderate telescopes were apparently overlooked by earlier observers using good telescopic equipment. Among such objects are the domes, very low blister-like swellings that occur almost exclusively on smooth mare-like terrain. They are well seen only when the solar illumination angle is quite small.

The older observers charted very few domes although they owned good telescopes in many cases. James Nasmyth went so far as to say that the dome north of the crater Birt was the only example of its kind on the Moon's visible surface. In 1932, Robert Barker drew attention to an especially good example inside the crater Darwin and in doing so seems to have stimulated the search for others. Nowadays, many are known. H.P. Wilkins and P.A. Moore alone found nearly 100. It is always possible that some of these may, in fact, be of relatively recent origin. Domes may be upliftings of the crust into blister-like swellings caused by gases or lava beneath the surface under pressure that is not quite sufficient to cause an actual volcanic eruption. There is no compelling reason to believe that mild lunar volcanic activity has entirely ceased.

On the other hand, V.A. Firsoff suggests that psychological factors may be responsible for the omission of domes from the older maps. An observer who is unprejudiced may miss seeing them if looking for a different type of object such as clefts, but they will be easy objects for anyone deliberately looking for them. Another possibility is changing interests and 'fashions' in lunar study. For example, authors of older books on the Moon continually

218

use the word 'mountain' in referring to lunar feaures. They speak of the mountainous terrain, the lunar ring mountains and the mountain ranges. A mind impressed by the mountainous grandeur of the Moon's surface and in awe of it is hardly likely to take much notice of mere domes, let alone take the trouble to record them. Because they were not recorded does not mean that they were not seen. Gruithuisen and Schröter saw domes on various parts of the lunar surface, but thought that they were clouds because they disappeared soon after sunrise. Hence, they didn't record them as they were thought to be impermanent features.

Clefts. H.P. Wilkins describes what may be a 'new' cleft on the floor of the great Alpine Valley. The Alpine Valley is a much observed part of the Moon and many hundreds of drawings and charts of it must be in existence. Very minute details have been recorded in the elaborate maps made of it. Along the bottom of the valley runs a cleft and another crosses the valley about one-third of its length from the western end. Wilkins relates that on April 20, 1953, he noted an easily seen delicate shadow-filled crack near the western end of the Alpine Valley – unfortunately, the aperture of the telescope he was using is not mentioned, but it was probably the 33-inch Meudon refractor. He says that the crack is quite prominent and wonders why it has not been previously recorded. Wilkins raises the question of whether cracking of this part of the Moon's surface is still going on.

According to Wilkins, a crack on the western part of the interior of the crater Reaumur appears to have lengthened since it was first charted by Goodacre. We do not know the exact date of Goodacre's observation, but the crack is shown in his lunar map published in 1910. Goodacre used an 18-inch reflector. Wilkins observed Reaumur with the 33-inch Meudon refractor on April 1, 1953, and noted that Goodacre's crack had a branch to it and that it extended to the south rampart. Wilkins remarks that it seems odd that Goodacre should not have shown the crack extending to the south rampart if, in fact, it really did extend so far in his time.

In the same crater, Dr S.M. Green, observing with a 12½-inch reflector, in 1939, saw a very delicate crater row within Reaumur, and, in 1953, Wilkins distinctly saw this not as a crater row, but as a cleft. There was no sign of anything like pits or depressions. It may be that the crater row has now widened into a true cleft. Since cracks and clefts are abundant in this area, it seems not improbable that new ones can still be formed.

Before leaving the subject of lunar clefts, it is interesting to note that the maps and charts constructed by different observers of the cleft systems associated with Triesnecker (Fig. 6.5), and of the interior of Gassendi (Fig. 6.6), show curious discrepancies that are difficult to explain. Again, this cannot be taken as evidence of change and the variations are most likely due to differences of visibility of individual clefts at different times attributable to illumination angle and libration effects and also the influence of seeing conditions, telescope aperture and interpretation of what is seen.

Conclusions. On the whole, the evidence for physical change on the Moon and the apparent appearance and disappearance of individual features

Fig. 6.5 *Triesnecker clefts as depicted by* (a) *P.B. Molesworth,* (b) *H.P. Wilkins (see also Fig. 4.18)*

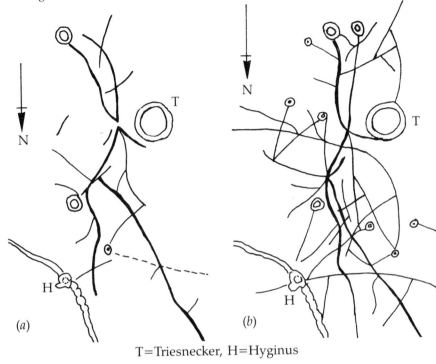

(a)   (b)

T=Triesnecker, H=Hyginus

Fig. 6.6 *Gassendi clefts.* (a) *Neison's chart,* (b) *Fauth's chart (see also Fig. 4.32)*

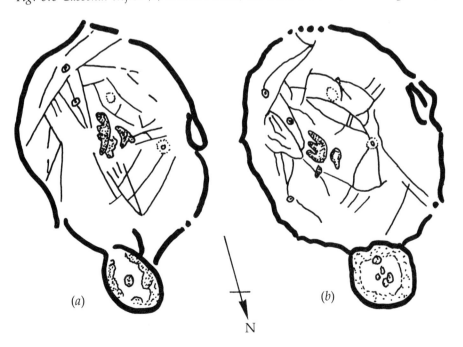

(a)   (b)

seems quite weak and largely attributable to errors of observation and interpretation of what is seen. Other possible contributory factors are the effects of libration, seeing conditions and telescope aperture, as mentioned above, and also the observers' experience and familiarity with the appearance of given areas of the Moon under a wide range of illumination and libration conditions. However, some of the phenomena may be due to temporary 'obscurations' on the Moon's surface that appear to be caused by localised clouds or mists; the curious variations in Hyginus N and the surrounding region may be explainable in this manner. The evidence for such phenomena is more convincing than that for actual physical change in lunar surface features, and will be the subject of the next section.

Clouds, mists and obscurations

Over the years, there have been frequent reports of localised phenomena on the Moon's surface in which the usual appearance or visibility of an object is temporarily altered while the surrounding topography retains its normal appearance. Hence, these phenomena cannot be ascribed to effects of our own atmosphere on the telescopic 'seeing'. Such sightings range all the way from temporary unaccountable fuzziness or unsharpness of features that are usually sharply defined to partial or complete obscuration of objects by vague mists ('unfocusable nebulosities') or distinct clouds.

Mare Crisium. The south part of the Mare Crisium is noted for these 'obscurations' and 'mists', some of them of quite long duration. On August 20, 1948, P.A. Moore was observing the mare, concentrating on the many hillocks and craterlets on its surface. On this occasion he was unable to see details on the west of the mare in the region of the Cape Agarum. To the west of the cape is a bay-like feature and this seemed to be filled with a mist that shone. Further to the east the crater Picard also seemed to be enveloped in mist.

Just to the east of the Cape Agarum and on the greyish floor of the Mare Crisium (Fig. 6.7) is a prominent group of craterlets, white spots and ridges which, owing to their arrangement, is often called the 'Trapezium' or 'Barker's Quadrangle' (Fig. 5.4), after R. Barker, who first gave a complete delineation of this feature in the 1930s. It is an easy object in a three-inch refractor and yet is not shown in the early maps of the Moon and Goodacre's map only hints at something there. It is strange that such a prominent feature has apparently been overlooked in the past. Could it be that for years the whole area was obscured by something like a mist? R.M. Baum once saw Barker's Quadrangle obscured by a cloud-like whitish patch.

Misty appearances have been seen on the Mare Crisium on several occasions, particularly in the Cape Agarum region. Perhaps something is being emitted from beneath the surface in this region such as fumes or vapor. An objection to this idea is that it is difficult to understand how mists or vapors could last for long in the virtually airless conditions on the Moon. They would be dispersed almost instantly in the near-vacuum. It could be that emissions of invisible gas stir up dust clouds and keep them moving

Fig. 6.7 Outline chart of the Mare Crisium showing the location of the 'Trapezium' (arrow)

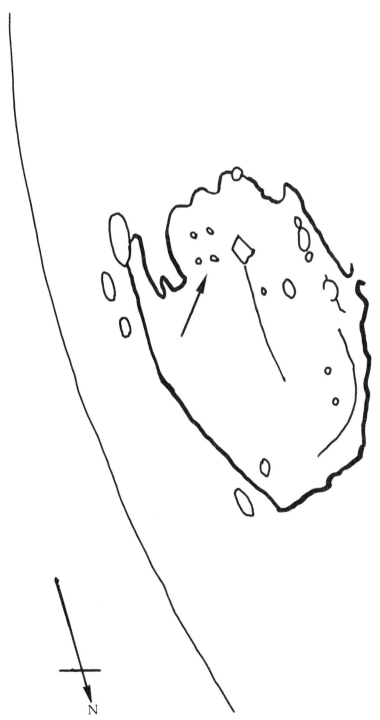

N

as long as the gas emissions last. Whatever else, it is difficult to explain obscurations and other similar phenomena in terms other than that of localised mists or clouds.

On the other hand, we must remember that the Mare Crisium lies close to the limb, and it is, therefore, fatally easy to be misled by odd lighting effects. A possible explanation of the absence of Barker's Quadrangle and other minute spots and craterlets from the older maps may be similar to the omission of domes previously mentioned; to an observer preoccupied with the grandeur of the lunar mountains, tiny white spots are not likely to be very interesting or worth recording.

West of the crater Picard in the Mare Crisium is a whitish spot that is said to be 'variable'. The majority of white spots on the Moon are actually craterlets that are too small to be clearly seen as craterlets. The white spot west of Picard appears not to be a craterlet; at least, that was W.R. Birt's opinion. It is an example of a type of feature seen on other parts of the Moon which because of their appearance in the telescope are called 'unfocusable nebulosities'. They appear to be surface deposits. Birt found that the spot west of Picard exhibited changes in brightness, and at times looked hazy. These phenomena have been seen in more recent times and P.A. Moore remarks that the Picard spot 'appears to be able to send out a certain amount of vapor'.

Some distance due north of Picard on the Mare Crisium is the smaller crater Peirce, and just to the north of Peirce is the still smaller crater Graham. H.P. Wilkins was unable to find Graham on May 12, 1927, when it should have been visible; it had its usual aspect on the day before and showed itself again, but only dimly, on May 13. Many years previously, W.R. Birt frequently found that Graham was quite invisible under lighting conditions in which it should have been easily seen. P.A. Moore saw this area 'misty grey and devoid of detail' on three occasions in 1948, while the surrounding region was clearly defined and sharp. This appearance was also seen by R.M. Baum on two occasions.

Plato. One of the most 'active' regions of the Moon with respect to obscuration phenomena is the floor of the crater Plato. Just after lunar sunrise and under good seeing conditions, four craterlets and sometimes other detail can usually be easily seen with moderate telescopes on the crater floor. Many more smaller ones are visible in larger instruments. From time to time, craterlets are difficult to see or impossible to find under conditions when they should be easily visible. Could it be that something is obscuring them at these times?

Elger, Goodacre and others noted that familiar floor detail in Plato was occasionally hidden by what appeared to be mist. Goodacre mentions observations of mists on the east side of the floor and at lunar sunrise Neison once saw the whole floor veiled by fog, which later disappeared. A.S. Williams was observing sunrise on Plato on March 27, 1882, and saw the floor 'glowing with a curious milky kind of light', so that on this occasion, the mist appeared to be self-luminous.

On April 3, 1952, H.P. Wilkins and P.A. Moore were observing Plato with the 33-inch Meudon refractor and both independently made drawings of it. These agree perfectly and in particular the well-known craterlet or pit at the inside base of the east wall is not shown in either. This object has been seen and drawn many times by different observers, but it was invisible on this occasion although other familiar floor detail further to the west was seen. On the same night, T.A. Cragg at the Mount Wilson Observatory, in California, could not see anything on the floor of Plato four hours after Wilkins and Moore made their drawings. This effect has been seen since in October, 1952, and April, 1953. The development of a mist in the east part of the crater interior and its spread over the rest of the floor in a matter of four hours is a conclusion difficult to resist.

In his book *Our Moon*, H.P. Wilkins relates how on the morning of August 12, 1944, he was observing the Moon with an 8½-inch reflector under good seeing conditions. Plato was under low evening illumination and about one quarter of its flat grey floor was already in shadow. On the illuminated part of the floor, all he could see was a bright round spot near the center. Wilkins identified this as being undoubtedly the largest of the floor craterlets but, understandably, was puzzled as to why a feature that usually is not easy to find should have taken on the appearance of a prominent bright spot. The craterlet appeared to be filled with a strongly reflective material; whatever it was, the usual aspect of the craterlet was completely changed by it.

Schickard. Like Plato, the interior of this great walled plain appears to be invaded by mists from time to time. On August 2, 1939, P.A. Moore saw Schickard filled by what looked like an extensive mist that obliterated most of the interior details. The mist even overflowed the lower parts of the crater wall. All Moore could see of the floor detail were two bright points.

On August 31, 1944, H.P. Wilkins was observing Schickard under fine seeing conditions. Instead of the usual distinct shadow-filled floor craterlets being easily visible, there were numerous bright white spots in the same places as the craterlets. On the next evening, the white spots were gone and the craterlets were seen in their usual places. This is very odd since the angle of illumination on the second evening was higher than on the first and the craterlets should therefore have been more distinct and shadow-filled on the first evening. The peculiar white spot appearance has been since seen by other observers. Together, these observations suggest that the floor craterlets in Schickard are the source of the obscuring medium, whatever it is.

Herodotus. The dark-floored crater Herodotus, the great winding valley–cleft named after Schröter that originates near its north wall and the brilliant crater Aristarchus to the west are among the most intensively observed formations on the moon's surface; anything of an unusual nature occurring here could hardly escape notice.

Some really convincing cloud-like phenomena have been seen in and near Herodotus. In 1893, W.H. Pickering was observing from Arequipa, in

Peru, and saw a peculiar appearance of the 'Cobra-Head', the wide enclosure near the beginning of Schröter's Valley, that was suggestive of an active volcano. He watched what was apparently a vapor column issuing from the Cobra-Head and made accurate drawings. Careful examination of these show great changes in the shape of the outline of the vapor column at different times that are detectable in a six-inch telescope. Pickering remarked that these changes seemed dependent on the angle of solar illumination since it was not until two days after lunar sunrise that any volcanic activity was noticeable.

On February 10, 1949, F.H. Thornton was observing the Cobra-Head with an 18-inch reflector when he saw what he described as 'a puff of whitish vapor obscuring details for miles around'.

The lunar Orbiter photographs clearly show that there is definitely no central mountain or any other elevation within Herodotus, yet, on July 15, 1955, V.A. Firsoff saw something closely mimicking a central mountain and casting a distinct shadow. Firsoff used a 6½-inch reflector at 200× and says that the central 'peak' was 'as plain as a pikestaff'.

J.C. Bartlett has seen a similar 'pseudo peak' in Herodotus and H.P. Wilkins plainly shows a white spot on the floor of Herodotus in a drawing made on March 30, 1950, using his 15¼-inch reflector. This drawing is reproduced in P.A. Moore's book *A Guide to the Moon* (First Edition).

Schröter's Valley has been seen apparently extending into Herodotus, which it certainly does not. Perhaps a cloud could also have been responsible for this appearance, which is not improbable in view of Pickering's and Thornton's observations.

Other clouds and mists. A 'temporary hill' three kilometers in diameter was seen south-east of the crater Ross D by T.A. Cragg, on July 16, 1964. Could these 'false hills' be due to clouds? Possibly in a similar category to the 'false hills' is a phenomenon described in a report by several witnesses that was published in *Scientific American*, in 1882, of 'two pyramidal luminous protuberances' on the Moon's limb. These appeared less substantial than the Moon's disc and slowly faded away. Note that these appeared at the limb, *not* the terminator.

On at least one occasion while observing sunrise on the crater Plato, I have seen with my eight-inch reflector dome-like objects on the crater floor in the same positions as some of the more prominent of the floor craterlets just as they emerged from the shadows. Ernst E. Both, who is curator of Astronomy at the Buffalo Museum of Science, Buffalo, New York, U.S.A, suggested to me that perhaps the craterlets stand on domes. However, this does not explain why the craterlets themselves were not visible and filled with shadow as they should have been. Maybe I was mistaken; however, T.G. Elger shows a strikingly similar appearance in a drawing of Plato at sunrise made on February 1, 1887. Not far from the center of the floor, which is about half covered with shadow, he shows a dome-like object similar to what I have seen, but it is about five times the size. There is no sign of the craterlet that normally occupies this position. The drawing is

225

reproduced in A. Mee's *Observational Astronomy*. Could these floor craterlets in Plato be emitting clouds that are rapidly dissipated by the Sun's heat?

On the evening of June 3, 1982, I was observing the craters Phocylides and Nasmyth with my eight-inch reflector. The seeing was fair and the Sun had not long risen on these formations. Most of the floor of Phocylides was in shadow, but the floor of Nasmyth was well lit. All I could see on the floor of Nasmyth was an ill-defined fuzzy hump near the center of the greyish floor. Outside Nasmyth and Phocylides several small craters were plainly visible, their interiors partly filled with shadow and all clearly defined. I was not familiar with the details on the floor of Nasmyth and made a drawing of the two formations without any idea of detecting any-thing of an unusual nature. When I compared my drawing with other drawings, charts and maps of Nasmyth, I could see that there was clearly something odd about the appearance of the object in Nasmyth as shown in my drawing. Some observers show a clearly defined crater in the same place as the hump, similar in appearance to the craters on the outside of the larger formations. Two drawings made by K.W. Abineri using an eight-inch reflector at higher and lower angles of illumination, respectively on October 28 and November 25, 1955, both show a crater near the center of Nasmyth and a crater is shown in drawings by F.H. Thornton and T. Maloney, both made on April 15, 1954; but in a drawing made using his eigh-inch reflector on April 21, 1891, T.G. Elger shows something that looks more like a hill than a crater if the shape of its shadow is anything to go on. Further, the drawing made by P.A. Moore with the aid of his 12½-inch reflector on July 7, 1955, in which the floor of Nasmyth is almost fully illuminated with only a trace of shadow under the west wall, shows abso-lutely nothing on the floor of Nasmyth, yet several small craters outside the wall are carefully drawn. A critic might argue that Moore simply omitted the central crater, but why would anyone drawing Nasmyth omit the central crater, which is a feature of Nasmyth, yet take the trouble to delineate carefully the small craters on the outside – assuming that the central crater was visible when Moore made his observation? The excellent photographs that I have examined of the area showing both morning and evening illumi-nation show the object in Nasmyth as a typical small crater.

The reader may think that I was mistaken about the hill-like appearance of the object on the floor of Nasmyth, but this does not explain why the similar-sized craters in the same general area had the usual appearance of clear-cut craters under these lighting conditions, which served as controls. Was something obscuring the crater in Nasmyth when I made my observa-tion? Whatever the answer, the odd appearance of this object is difficult to explain.

In his book *Strange World of the Moon*, V.A. Firsoff, the noted lunar and planetary observer, mentions low domes in the craters Posidonius and Charcornac. In his observational drawing reproduced in the same book, five domes are shown in Posidonius, and three in Charcornac. Strangely, no one else, to my knowledge, mentions domes in these craters or appears to have seen them. I have observed Posidonius closely and have never seen any. It is just possible that Firsoff may have misinterpreted the bright halo around the craterlet between the central crater and the south wall of Posidonius as a dome and the broad low mounds just within the south wall may also have been misconstrued; but it is difficult to explain the remaining two domes and the three in Chacornac because they do not exist. What, then, did Firsoff see? Clouds? The central crater in Posidonius was said by Schröter to be subject to obscuration and the French astronomer, Lamech, mentions that one of the craters in the south part of the floor is likewise subject to obscuration.

Another French astronomer, Gaudibert, saw the floor of the Mare Nectaris so foggy on January 18, 1880, that almost no detail could be seen except the crater Ross. The fog reached half way into the crater Fracastorius. Yet another French astronomer, Charbonneaux, observing with the 33-inch Meudon refractor, in 1902, observed a small but quite definite cloud form near to the crater Thaetetus. Clouds have been suspected in Conon, a crater situated in the lunar Apennines, and ill-defined cloud-like objects and mistiness have been seen in the Sinus Iridum.

In one part of the lunar Alps is a region that varies in brightness, and at times some of the mountain peaks appear poorly defined although the surrounding area looks sharp and clear.

Barcroft and others have often seen mists in the crater Timocharis on the Mare Imbrium and mistiness was seen frequently in Tycho by W.R. Birt during the decade commencing 1870. Barcroft saw the floor of Tycho 'strangely ill-defined' on occasions, and T.G. Elger remarked on the indistinctness of the crater floor, as already noted in Chapter 4. The central mountain of Tycho was seen by R. Barker, on March 27, 1931, to be of a strange grey tint, yet the crater interior was filled with shadow. W. Haas saw a 'milky luminosity' on the east wall of Tycho within the shadow, on July 14, 1940. Another example of a self-luminous mist was seen by E.E. Barnard, at the Lick Observatory, in 1892, when the interior of the crater Thales was filled with a pale luminous haze.

P.A. Cattermole reported the apparent disappearance of the central mountains in the crater Eratosthenes, on May 11, 1954, although the surrounding features were clearly visible.

Variable light and dark spots and streaks

Plato. Mädler described four parallel light bands on the floor of Plato running in a north–south direction (Fig. 6.8). According to Wilkins, these bands no longer exist. V.A. Firsoff says that he has seen hints of four dark bands on the crater floor with the aid of a red filter, but thinks that he may have been mistaken. Mädler's light bands, therefore, seem to have decreased in intensity or even to have disappeared since his time.

The floor of Plato is criss-crossed with a complex pattern of very pale light streaks, and these seem to exhibit changes in their extent and visibility; among the many charts that have been made of them by various observers in the past, there are curious inconsistencies. The pattern of streaks 'develops' as the solar illumination angle increases and appears to vary from lunation to lunation. Some of these streaks connect the bright spots, while others are simply extensions of the streaks originating from the crater Anaxagoras many miles away north of the Mare Frigoris. The south-east part of the floor is occupied by a diffuse triangular light area called the 'Sector', whose apex touches the nearly central craterlet (Fig. 6.9).

Eratosthenes. Perhaps the most notable example of variable dark spots and streaks are those associated with the crater Eratosthenes. As the Sun rises higher and higher, the outlines of the crater become indistinct and finally become almost lost in a complex pattern of light and dark markings. These seem to move around and change position in a remarkable manner. This does not mean that they can actually be seen moving as one looks at them, but if careful drawings are made on consecutive nights over several luna-tions, comparison of them will show discrepancies indicating that certain details of the pattern of light and dark markings change in a way that is not attributable to the Sun's elevation or to the libration. These 'pseudo-shadows', as the dark markings have been called, do not coincide with the crater walls or other relief features, especially on the west side.

W. Pickering made an especially painstaking study of these high Sun markings in Eratosthenes. He observed from Arequipa, in Peru, under excellent seeing conditions, and found that the general pattern of dark markings goes through a predictable cycle in each lunation, but certain small details showed unpredictable positional changes from one lunation to another; some of the markings would move up the inner slopes of Eratosthenes and cross over ridges in a manner and with speeds that reminded him of slow-moving swarms of insects. He thus was led to pro-

Fig. 6.8 Light streaks on the floor of Plato according to Mädler

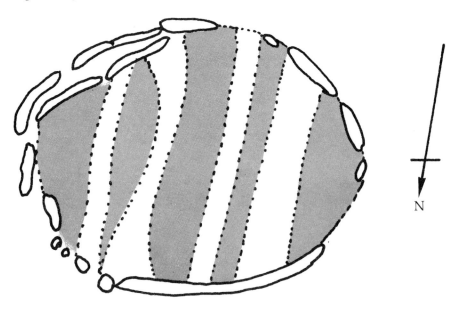

N

pose his theory of 'lunar insects' that, unfortunately, exposed his observations to severe and undeserved criticism. Pickering was undeniably a skilled and excellent observer, but his interpretations of some of his observations may have been too speculative.

In more recent times, a detailed study of the 'moving bands' of Eratosthenes was undertaken by Inez Beck, of Wadsworth, Ohio, using a six-inch reflector. From an examination of her file of over 150 observational drawings of Eratosthenes, it emerged that there are distinct differences in the pattern of light and dark markings at nearly identical colongitudes in different lunations and that it is difficult to predict what will happen next. Apparently, this is the first detailed study of the changes of aspect undergone by Eratosthenes since Pickering's work and it confirms his findings. This information was taken from an illustrated article about Beck's work on Eratosthenes that was published in Volume 22 (1970) of *The Strolling Astronomer*. The shifting pattern of dark markings within Eratosthenes is reminiscent of the 'pseudo shadows' in Bullialdus, mentioned in Chapter 4. Variable high Sun markings have also been observed in the crater Copernicus.

Alphonsus. On the floor of the crater Alphonsus are several dark spots at least some of which are suspected of 'variability'; certainly there are curious discrepancies between the delineations of different observers (Fig. 6.10). The three most prominent spots form a triangle, two on the west part of the floor and one on the east. Of the other spots, one lies at the foot of the west wall between the other two on the west, closer to the more northerly

Fig. 6.9 Light spots and streaks on the floor of Plato, 1879–82. (Redrawn after A.S. Williams, slightly simplified)

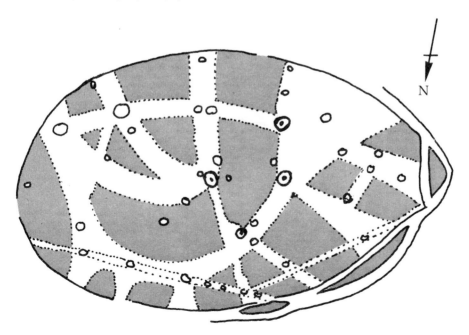

N

spot, and just to the north of the more northerly spot lie two others quite close together. The remaining spots are quite small. One lies approximately between the large eastern spot and the central mountain, and the other is situated at the foot of the south-west wall (Fig. 6.11). These spots are most prominent near local noon and are almost invisible at sunrise and sunset.

The apparent darkening of the spots with increasing solar illumination angle is illusory. At sunrise, the small irregularities in the surrounding terrain cast long shadows and the spots appear to be less dark by comparison. As the shadows disappear with increasing solar altitude, the spots actually become brighter but subjectively seem to darken because of a contrast effect and appear darkest at full Moon.

*Fig. 6.10 Dark spots in Alphonsus according to various authors (see also Fig. 6.11) (Redrawn from Goodacre, W. JBAA **41**(8), 380 (1931))*

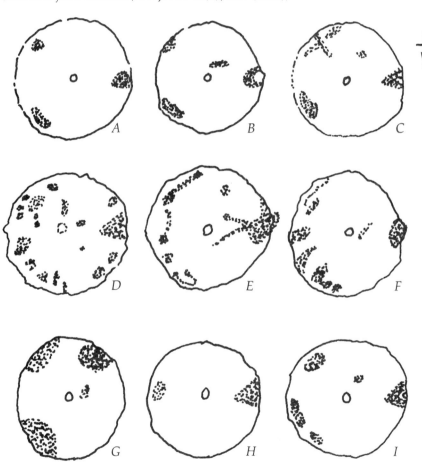

A,B,C. Pickering's photographic atlas
D. Fauth   G. Schmidt
E. Pickering   H. Neison, Elger
F. Lohrmann   I. Mount Wilson 100 inch photograph

Mädler, Webb and Neison considered that the spots were smooth level areas, but Klein and Williams saw them in their true aspect as slight depressions with rough surfaces. Klein also demonstrated that the apparent changes in intensity of the spots were essentially dependent on the solar illumination angle, but W.H. Pickering concluded that the intensity changes were independent of the Sun's altitude. He considered them to be patches of vegetation.

Three of the spots near the west wall have craterlets near their centers and a winding cleft connects the spots.

Atlas. The crater Atlas has two well-known dark spots on its floor, one on the north-west, and the other at the extreme south end (Fig. 6.12). T.G. Elger first noticed the variability of the south spot in 1870–1, and, in 1883–7, A.S. Williams made observations that enabled him to construct a light curve for both of them. W.H. Pickering gave much attention to the spots and confirmed their variability. Pickering found that within the south spot are darker areas and also that there are several craterlets close by and one near the center of the south spot. He also found a craterlet in the center of the north spot and some delicate clefts. The dark spots in Atlas and the craterlets associated with them, therefore, resemble the Alphonsus spots.

Fig. 6.11 Dark spots in Alphonsus. November 22, 1980. Colong. 97.4°. F.W. Price. Eight-inch reflector

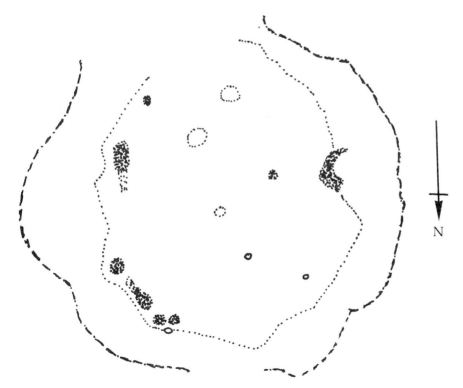

N

It has been suggested that the dark-colored spot material in these craters is volcanic ejecta and that variations in their appearance that are not attributable to the solar illumination angle may be due to some kind of continuing volcanic or other igneous activity.

Other variable spots. On the floor of the crater Endymion there are patches that are greyer than the rest of the floor. They change shape as the Sun rises higher, some enlarging, others growing smaller and yet others disappearing. W.H. Pickering again considered these to be vegetation, but the generally held view is that the variable patches are caused by the Sun's heat causing changes in tint in some unusual material on the surface of the crater floor. The matter is still undecided and is an open question.

Fig. 6.12 *The dark spots in Atlas. Appearance five days before full Moon. December 17, 1961. F.W. Price. Three-inch refractor*

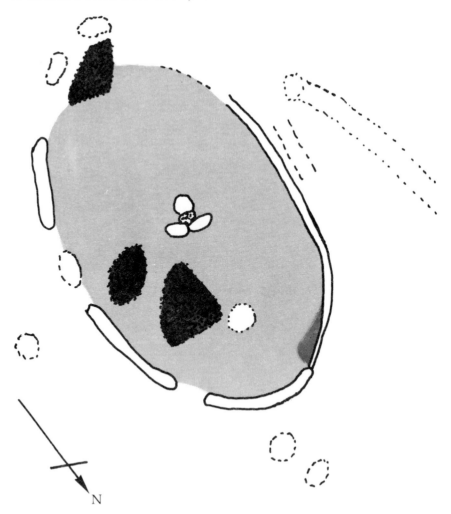

N

H.P. Wilkins has noted variable dark spots on the floor of Grimaldi. There was once a dark area on the floor of Petavius that was seen by all the earlier observers. Today, it can no longer be traced.

On the floor of the crater Werner is a spot that Beer and Mädler declared to be as bright as any part of the lunar surface, and they estimated its brightness as equal to that of Aristarchus. The spot is still quite bright, but nowhere near as bright as Aristarchus. It has definitely faded since Beer and Mädler's time.

Banded craters. The lunar crater Aristarchus is noted not only for its great brilliance, but also for the dusky radial bands that run up the interior slopes of the east wall (Fig. 6.13). There are several of them and two or three of the most prominent are easily visible in a three-inch refractor.

*Fig. 6.13 The dusky bands in Aristarchus (full moon aspect). Redrawn after R.A. McIntosh. (JBAA, **71**(8), 387 (1961))*

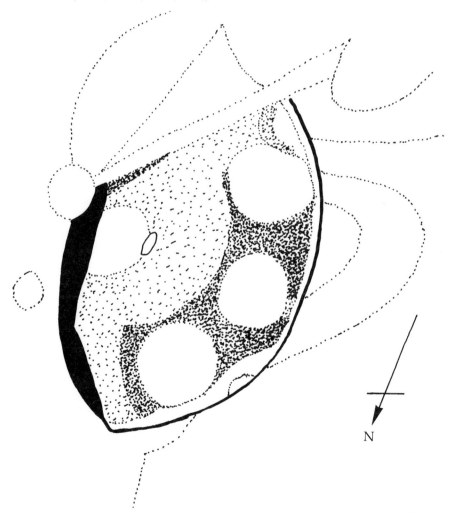

N

When the Sun has risen sufficiently to expose the inner east wall of Aristarchus, faint grey marks can be made out which 'develop' into bands and become more prominent as the Sun climbs higher. They are most prominent and numerous under a high Sun, and even 'spill over' the east wall onto the surrounding surface adjacent to Aristarchus. They gradually fade away as the sun sets.

Large telescopes reveal that the bands in Aristarchus are not continuous but have 'fine structure' made up of dots and dashes, as was first established by H.P. Wilkins with the 33-inch Meudon refractor. The bands start some distance from the central elevation and Favarger has seen brilliant star-like points between them, but P.A. Moore and R.M. Baum see them as 'bright patches'.

The radial bands are quite prominent and yet none of the older seleno-graphers including Schröter, Beer, Mädler and Lohrmann seem to have seen or recorded them, as mentioned in the previous chapter. Schmidt made a large scale drawing of Aristarchus and even he does not show the radial bands. It was not until 1868 that they were first shown in a published drawing by J. Phillips; however, a drawing made by Lord Rosse five years earlier shows them, as was also previously mentioned. There is a lengthy account of Aristarchus in Neison's book, but he does not say anything about the radial bands and it was not until 1884 that two of them were recorded by Sheldon. From that time onwards they seem to have become more prominent and recent charts show up to nine bands.

R. Barker concludes from his investigations that the Aristarchus bands have increased in prominence during the last 100 years or so. If this be denied, then it is difficult to account for them having been apparently 'missed' by Schröter, Mädler, Lohrmann and Schmidt. All four of them carefully studied Aristarchus under all solar illumination angles.

For many years, Aristarchus was thought to be the only 'banded' crater, but then others began to be discovered. East of the crater Bullialdus is the small crater Moore, named after P.A. Moore, who discovered bands on its interior in 1949. Another example is the crater Birt situated near to the straight wall. More than 100 small banded craters are now known. Interest-ingly, Aristarchus is the only example of a large banded crater.

What are the bands? None of the explanations so far put forward to explain their nature and origin is entirely satisfactory. One idea was put forward by P.A. Moore, who suggested that the Aristarchus bands are formed of a lowly type of vegetation. He says that there could be fine radial clefts within Aristarchus coinciding with the bands. Gases are emitted from these clefts that support the vegetation as the sun rises higher in the sky and causes the vegetation to develop; as we shall see later, gas emission from Aristarchus has been established. Two observations support Moore's idea. First, the fact that the bands develop from the center of the crater and spread outwards would be anticipated of the behaviour of vegetation since most gas would be expected to be emitted from near the central mountain, which is the activity center, and less would be emitted further out. Second, scrutiny with large telescopes reveals that the bands have a darker strip at the center whereas the parts of the band on either side are paler and more fuzzy. This is what would be expected on Moore's theory because the vegetation would probably show its densest growth close to the clefts from where the life-supporting gases are emitted. The gas could be the heavy carbon dioxide which is the last sign of volcanic activity that is dying down.

Lights, flashes and glows

Active volcanoes? We have already noted that bright craters like Aristarchus can look so bright on the dark part of the Moon during the waxing phase that Sir W. Herschel and also Schröter thought they were witnessing volcanic eruptions. V.A. Firsoff remarks that he once saw Aristarchus 'like a star through a cloud that completely obliterated the rest of the Moon'. He cautions inexperienced observers against being misled by star-like ghost images caused by reflections from the lens surfaces in the telescope eyepiece being projected onto the dark side of the Moon. Herschel is unlikely to have been so misled because he was very careful about the location of these bright spots on the lunar surface.

Among other similar phenomena seen by Herschel were three bright points of light on the Moon's dark side in March 1787. On April 19, he describes 'three volcanoes', the brightest of which was about 3' 57" from the north limb. This looked to him like a 'small piece of burning charcoal covered by a very thin coat of white ashes seen in faint daylight'. He saw them again on the following night and the brightest of the 'volcanoes' was even brighter. During the total lunar eclipse of October 22, 1790, he saw about 150 very luminous spots scattered over the Moon's surface.

Favarger saw bright points of light like stars on the interior east slopes of Aristarchus; other observers have seen similar bright points, but on this occasion they were unusually bright. On three nights during May 1821, Ward and Baily observed a bright spot near Aristarchus on the dark part of the Moon.

Schröter once saw a tiny point of light on the Moon's dark side near the crater Agrippa and in 1788, on September 26, he saw a star-like spot shining out in the Earthshine in the Alps south-west of Plato, which continued for 15 minutes on this occasion. Could it have been a volcanic eruption?

In the *Astronomical Register*, Volume 5, p.114, T.G. Elger reports seeing a light spot of about seventh magnitude on the dark side of the Moon. This lasted for an hour and grew fainter during the final 15 minutes. After another half hour, it had gone. Elger mentions that he had seen similar phenomena many times previously.

Bright spots of a strange fiery color have been seen near the crater Plato, also in the Teneriffe Mountains and near to Mont Blanc, at different times during the course of more than a century.

P.A. Moore saw three brilliant star-like points of light on the wall of Darwin on October 19, 1945.

H.P. Wilkins describes a curious observation he made in connection with the isolated lunar mountain LaHire on November 28, 1922. On this occasion, the mountain was casting its usual long shadow, but what was unusual was that this time the shadow was seen cut through by a white streak that disappeared after 20 minutes. If it was a cloud hovering over the shadow, why did it not itself cast a shadow on the surface beneath it? None was seen by Wilkins. Alternatively, if it was a surface feature, why was it not hidden by the shadow of LaHire? Could it have been a glowing lava flow that suddenly appeared from below the surface?

Meteorite impacts and a lunar atmosphere? The glowing points of light seen on the dark side of the Moon described in the previous section were all quite long-lived, lasting for several minutes at least. Similar lights have been seen lasting only a few seconds or less. An example was the tiny luminous speck that W. Haas saw near the crater Gassendi, on July 10, 1941. The speck moved across the surface toward the crater wall and lasted for only a second. H.P. Wilkins saw a bright speck inside Gassendi on May 17, 1951, that lasted for a second and left an afterglow for two or more seconds. Another instance was seen by A.W. Vince, who was observing the Earthlit part of the Moon with a 6⅜-inch refractor on the evening of April 15, 1948. He saw a sudden flash on the dark limb to the north of Grimaldi which was visible as a dusky patch. The flash was only momentary and was estimated to be about as bright as a naked-eye star of the third magnitude.

A bright flash lasting for three seconds was seen on the dark part of the Moon by A.J. Woodward, on August 8, 1948. He said that the flash changed color from bluish-white to greyish-yellow.

The Japanese astronomer Saheki saw a yellowish-white stationary flash lasting a quarter of a second on August 25, 1950.

Maybe the best known example is F.H. Thornton's sighting of a tiny but brilliant flash just inside the rim of the west wall of Plato. It was of an orange-yellow color, and was described by Thornton as resembling 'that flash of an AA shell exploding in the air at a distance of about ten miles'. At first, he thought that it might have been due to a fall of rock, but rejected this interpretation upon realising that the flash must have been over half a mile away from the foot of the crater wall. Thornton was using a nine-inch reflector under excellent seeing conditions. Can these flashes be due to lunar meteors?

The difficulty in interpreting these short-lived light specks as meteors is that an appreciable atmosphere is required to provide the frictional heat needed to cause a meteor to burn up and emit light, and we are told that the Moon is airless. On the other hand, the flash may be due to the impact of the meteor on the surface. The question of the existence of a lunar atmosphere is a vexed one but how else are we to explain an apparent lunar twilight effect seen by H.P. Wilkins on the evening previous to Vince's observation? Wilkins was observing the crescent Moon with his 12-inch reflector. The evening was very clear and the Moon's Earthlit part was plainly seen. The peaks of the Leibnitz Mountains were brilliantly visible on the darkened south limb and were seen to be connected by very delicate but quite distinct bright threads of light which gave the appearance of the southern cusp of the crescent being unusually prolonged. If this was not an effect caused by refraction in the lunar atmosphere, what else could it have been? W.S. Franks witnessed a similar apparent prolongation of the south cusp of the crescent Moon along the Leibnitz Mountains, on March 20, 1912. He was observing with a good quality six-inch refractor. The effect appeared as a faint line of light projecting well into the dark hemisphere. There is no doubt that the effect is genuine as other observers have seen it including P.A. Moore, Vaughn, Barcroft and W. Haas. Even Mädler saw it. The phenomenon was apparently first seen by Schröter, who often found the horns of the lunar crescent to be prolonged as a light ring around the dark part of the Moon. What is odd is that the twilight is not always seen at every crescent phase.

Transient lunar phenomena (TLP). The term 'transient lunar phenomena', or TLP for short, can, of course, be correctly applied to any of the temporary clouds, obscurations, mists and flashes of light just described. However, the term is now used almost always for short-lived reddish glows that are seen from time to time on the Moon's surface, although there may also be white or even bluish glows. Observation of such events goes back many years. Mädler mentions seeing several times, between 1830 and 1840, a reddish appearance of the lunar surface closely west of the crater Lichtenberg, and an American observer, Barcroft, saw a reddish-brown color in this area on October 18, 1940. It has also been seen by Haas and R.M. Baum, the latter seeing a reddish glow in the region on January 21, 1952. It is possible that the coloration may be due to the Sun's rays striking a peculiar surface material at a critical angle. A small area glowing red was seen by Valier, on the dark side of the Moon, on May 19, 1912, and Maw saw a 'small reddish' spot inside the crater South. Both Goodacre and Molesworth saw a bluish 'glare' in Aristarchus in 1931, and on February 22 of that year, a reddish glow was seen in the same crater by Joulia. Certain authorities doubted the reliability of these sightings, however.

The lunar crater Alphonsus occupies a special place in the literature of lunar surface happenings because something that was observed within its walls during November, 1958, was the only really well authenticated example of a TLP at that time. Local events had been reported inside Alphonsus previous to this. As far back as 1882, Klein asserted that there was volcanic activity inside Alphonsus, but the reliability of this report may be questionable. A more recent report is the photographic observation made by D. Alter, in 1955. He photographed Alphonsus and Arzachel with the 60-inch Mount Wilson reflector in both infra-red and ultra-violet light. As is well known, infra-red wavelengths penetrate haze and mist whereas ultra-violet is scattered by it. On many occasions, Alter found that in violet light pictures, the clefts on the west part of the floor of Alphonsus looked blurred, whereas this effect was not seen in the red light pictures. Alter cautiously suggested that the blurring effect in the violet light pictures could be due to light scattering by a temporary or permanent atmosphere on the floor of Alphonsus, possibly an emission of gas from the clefts. Then, on the night of November 3/4, 1958, the Russian astronomer N.A. Kozyrev was using the 50-inch reflector of the Crimean Astrophysical Observatory to take spectrograms of Alphonsus. He was doing this as a result of a suggestion by P.A. Moore to him and other observers that they keep Alphonsus under close scrutiny. Because the 50-inch reflector does not have a guide telescope, Kozyrev had to watch through the main telescope and guide it during the photographic exposure. On this occasion, the spectrograph slit was centered on the central mountain of Alphonsus. As he watched, Kozyrev noticed a blurring of the central peak of Alphonsus at 01.00 UT, which seemed to have become enveloped in a reddish fog. Later, at around 03.00, the central peak became unusually bright and suddenly started to fade. By about 03.45, Alphonsus resumed its normal appearance. During

Fig. 6.14 Chart of Lunar Transient Phenomena. (From data of B. Middlehurst and P.A. Moore). Redrawn from New Guide to the Moon *by P.A. Moore, Norton, 1976 by permission. Copyright © by P.A. Moore*

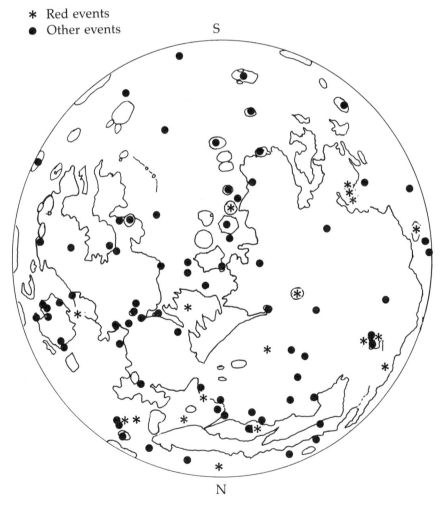

* Red events
• Other events

this time, Kozyrev had exposed several photographic plates and that taken between 03.00 and 03.30 was uncommonly interesting. This and the others indicated that hot carbon gas had been ejected, which caused the temperature to rise to 2000°C. Although this conclusion was doubted by some other theorists, there is no doubt that some kind of disturbance had occurred in Alphonsus and that Kozyrev had witnessed a TLP.

In the weeks following Kozyrev's observation, the central peak of Alphonsus was studied by several observers to ascertain whether any permanent change was visible. There were many reports of red patches in the area and these were attributed to pigmented matter being ejected when the disturbance occurred. G. Kuiper, using the 82-inch McDonald reflector, saw no structural changes or unusual color in the central peak. The central peak of Alphonsus was studied by M. Focas with the 24-inch refractor at the Pic du Midi observatory, on November 19, 1958, and he too saw nothing unusual about it, yet, on the very same evening, Poppendick and Bond reported seeing a 'diffuse cloud over the central mountain'. Also, at the same time Focas was studying Alphonsus, G.A. Hole was observing it with his own 24-inch reflector. Although he saw no structural changes, he did see a reddish-brown tinge on the south slopes of the central peak and satisfied himelf that this was not due to instrumental effects or atmospheric refraction. He saw the reddish patch again on December 19, 1958. What seems strange is that the colored patch was apparently not seen by the other observers with their large telescopes. Was the reddish patch seen by Hole and other observers, including H.P. Wilkins, really there before Kozyrev's observation on November 3, 1958?

Other red TLP in Alphonsus have since been reported and Kozyrev himself made another spectrographic observation on October 23, 1959, indicating some kind of activity although nothing was seen visually. In connection with this it may be significant that not long afterwards, on January 6, 1960, B. Warner saw a red patch in Alphonsus. Today, the red patch seems to be no longer visible.

Not long after, in 1963, two observers at the Lowell Observatory at Flagstaff, Arizona, J. Greenacre and E. Barr, saw definite red and pink color phenomena on the lunar surface near Aristarchus on October 30. Similar color effects were seen in the next month and verified by P. Boyce using the 69-inch reflector at the Perkins Observatory. These observations served to confirm the reality of these 'red events'. Spectrographic observations have since confirmed gaseous emissions from Aristarchus.

Although 'red events' may sometimes be seen directly in the telescope, they are more easily visible and faint ones more certainly detected by an optical device that enhances the contrast of the color effects. This is done by alternate use of red and blue filters placed just in front of the telescope eyepiece. The device will be described in detail in the next chapter.

During the three years following the Greenacre and Barr observations, sightings of TLPs were reported by different observers. The most notable of these was the 'red event' in the crater Gassendi, on April 30, 1966, which was detected by P.K. Sartory with the red and blue filter device mentioned above. It was seen and verified independently by many observers. The main feature appeared as an unmistakable orange-red wedge-shaped area on the crater floor extending from the walls to the central peak. P.A. Moore was one of those who saw it and he described the phenomenon as 'the most unmistakable red event that I have ever seen on the Moon'. It lasted for about four hours.

Quite independently of each other, P.A. Moore at Armagh, in Ireland, and Barbara Middlehurst and colleagues, of the University of Arizona and Goddard Space Flight Center in the USA, commenced putting together a catalog of all reported unusual happenings on the Moon since scientific telescopic study of the Moon began. So similar were the results that Moore and Middlehurst jointly published the catalog. Later, in 1971, Moore enlarged the catalog and brought the list of lunar 'happenings' up to 713.

An interesting fact that comes from an analysis of the events is that they are not randomly distributed. There is a definite tendency for 'red events' to 'cluster' around the 'shores' of circular maria and in areas where there are rills. The borders of the Maria Crisium, Serenitatis and Imbrium are especially 'event-prone'. Further, 'red events' are mostly seen in the eastern hemisphere of the Moon. In 1963, J. Green showed that the frequency of TLP was greatest at lunar perigee, when the Moon is closest to the Earth. At such time, the Moon's crust is under greatest gravitational stress. Confirmation of this was later forthcoming from Barbara Middlehurst and co-workers. This finding fits in well with the idea that TLP may be caused by or be related to emissions of gas from beneath the lunar crust; the concentrations of TLP-prone areas where rills abound is also consistent with this. Gas emission from beneath the lunar surface has been definitely shown to occur, as revealed by the spectrographic studies of Aristarchus and similar ones of Grimaldi. A. Mills suggested that the gas emissions cause solid particles to move around and the friction caused by this produces the glows.

There is also a definite connection between the TLP and Moonquakes. Seismic instruments were set up on the Moon's surface by the Apollos 12 and 14–17 astronauts. These instruments showed that Moonquakes were frequent, but feeble by Earthly standards. They are genuine and produced by causes similar to terrestrial Earthquakes. The distribution of 'Moonquakes' is plainly similar to the pattern of 'clustering' of the TLP areas. Moonquakes are most frequent near lunar perigee as predicted and so all these correlations seem to add up to a consistent picture. There does not seem to be any correlation of TLP frequency with the 11-year Sunspot cycle.

Light shed on lunar 'mysteries' by orbiting space craft

The Linné controversy was settled once and for all by Orbiter close-up photography. Linné was revealed to be a perfectly ordinary craterlet with a 'fresh' appearance. It does not show any signs of having once been a larger crater whose walls collapsed inwardly, as many have believed in the past to have been the cause of the supposed 'disappearance' of Linné. It has undoubtedly always looked as it does today and has definitely not undergone physical change of any great magnitude. Orbiter photography confirms the presence of many small craterlets on the floor of Plato seen in Earth-based telescopes and many other minute craterlets beyond the reach of most telescopes. However, no light is shed on the strange way in which some of the craterlets 'come and go' as mentioned earlier, a phenomenon that does not appear to be related to seeing conditions or illumination angle. The only explanation for this would be the obscuring effect of mists or other exhalations that occasionally form on the crater floor.

The actual number and configurations of the clefts associated with Triesnecker and on the floor of Gassendi are quite unarguably revealed by Orbiter photography. The discrepancies between the drawings of these cleft systems by different observers is no longer a lunar mystery and must lie in differences of interpretation of what is seen, librational effects, seeing and illumination angle under which the observations by different observers were made and possibly also be due to physiological and psychological factors.

The true nature of the dark radial bands on the inner east wall of Aristarchus is shown by Orbiter and Apollo photography. The bands are simply radial valleys of darker material than the higher land between. What are still mysteries are why they seem to have been overlooked by earlier observers and their peculiar visibility characteristics during the course of a lunation.

That there is no central elevation on the floor of the nearby formation Herodotus is plainly shown by Orbiter photography, as mentioned earlier. The enigmatic 'pseudo peak' that has been seen there at times must therefore be due to some kind of transient phenomenon such as a cloud and not in this case to errors of interpretation by the observers who have seen it.

In at least one instance a transient phenomenon was seen by the astronauts aboard the Apollo 11 space craft on July 19, 1969. Using binoculars, Collins, Armstrong and Aldrin stated that they saw luminosity of the north-west wall of Aristarchus 'more active' than other parts of the surface. At about the same time, Earth-based telescopic observers confirmed the presence of a TLP in Aristarchus. Mistiness seems to be present in some of the Lunar Orbiter V pictures of Aristarchus, especially frame HR 194–3. Valuable information about the possible origin of at least some TLP was obtained by the Apollo 15 space craft in 1971. The command module was equipped with an alpha particle detector. Alpha particles are released by decay of the unstable atoms of the gas radon, itself a product of radioactive decay of uranium and thorium. Diffusion of radon through the Moon's outermost surface layer into the sparse lunar atmosphere would therefore be accompanied by alpha particle emission. When the Apollo 15 space craft passed over Aristarchus at a height of seventy miles, the alpha particle detector indicated an increase in the rate of alpha particle emission from the radon isotope radon-222 but not from radon-220. This shows that the emission was not produced by a localised surface excess of thorium or uranium but must have come from diffusion of radon-222 from beneath the surface. Therefore, radioactivity in the Aristarchus region must have been the cause. It was concluded that the emission of radon is related to the internal processes that occasionally give rise to expulsion of gases to such an extent that their effects are visible in Earth-based telescopes. This plainly correlates with TLP and it is interesting to note that in another 'event-prone' area (Grimaldi), similar radon emissions have also been detected.

In conclusion it may be said that controversies over the physical presence or absence of this or that surface feature, or over the precise configurations of surface detail, especially at the limit of telescopic resolution, have been settled once and for all by the Orbiter close-up photography. Theories about whether changes have occurred, such as the 'disappearance' of Linné, are now obsolete. It was indeed fortunate that at the time of the Apollo flight, a TLP was observed in Aristarchus. However, the long term 'variability' of intensity and extent of dark and light spots or streaks and the unpredictable TLP can only be practically studied by Earth-based observation. For obvious economic and practical reasons, prolonged surveillance of the Moon's surface from orbiting spacecraft for these types of phenomena is out of the question. Earth-based study of TLP and long-term variability is therefore an important area of lunar research in the post-Lunar Orbiter era.

Further reading

Books

Strange World of the Moon. Firsoff, V.A. Hutchinson and Co., London, England (1959). Chapter 8.

The Old Moon and the New. Firsoff, V.A. Sidgwick and Jackson, London, England (1969). Chapter 13.

Our Moon. Wilkins, H.P. Frederick Muller Ltd, London, England (1954). Chapters 11 and 14.

Papers and Articles

'Does Anything Ever Happen on the Moon?' Haas, W.H. *Journal of the Royal Astronomical Society of Canada*, **36**, (6) 237–72; (7) 317–28; (8) 361–76; (9) 397–408 (1942).

'Lunar Notes.' (Drawings by Inez Beck and notes of the variable dark markings in Eratosthenes.) *The Strolling Astronomer*, **22**, (9–10) 175–77 (1970).

'Linné.' Wilkins, H.P. *Journal of the British Astronomical Association (JBAA)*, **64**, 86–8 (1953–4).

'A Note on the Lunar Formation Linné Expressing Some Evidence in Favor of the Rejection of the Concept of Change There.' Baum, R.M. *The Strolling Astronomer*, **10** (7–8), 93–4 (1956).

'A Further Note on Linné.' Moore, P.A. *The Strolling Astronomer*, **18**, (1–2) 25–7 (1964).

'Centenary Reflections on Linné.' MacDonald, T.L. *JBAA*, **77**, 39–41 (1966/7).

'The Linné Affair.' Robinson, J.H. *JBAA*, **81**, 34–6 (1976/7).

'The Linné Controversy.' Moore, P.A. *JBAA*, **87**, 363–8 (1976/7).

'Maedler's Square – An Autopsy.' Bartlett, J.C. *The Strolling Astronomer*, **6** (8), 122–35 (1952).

'Maedler's Square – A Study in Lunar Paradox.' Bartlett, J.C. *The Strolling Astronomer*, **4** (12), 1–13 (1950).

'Maedler's Square – An Alternate Interpretation.' Moore, P.A. *The Strolling Astronomer*, **5** (7), 3–7 (1951).

'The Picard–Cape Agarum Area of the Moon.' Moore, P.A. *JBAA*, **59**, 250–2 (1948/9).

'Variations Within Plato.' Wilkins, H.P. *JBAA*, **53**, 190–2 (1942/3).

'Physical Changes on the Moon.' Barker, R. *JBAA*, **48**, 347–53 (1937/8).

'Lunar Changes.' Pickering, W.H. *7th Report of the Section for the Observation of the Moon*. British Astronomical Association (BAA), London, England (1916).

For references to transient lunar phenomena (TLP) see the list of further reading at the end of Chapter 7.

7

Suggestions for Research

I sincerely hope that the reader who has come thus far will not be content merely to read about the Moon but will want to do systematic practical work or research. The word 'research' is somewhat ambiguous insofar as it means different things to different people. In a narrow sense, scientific research may be defined as investigational work that results in new facts and definite contributions to scientific knowledge that are publishable in the standard refereed scientific journals. In a broader sense it may mean any kind of investigational work that yields data, even if these are not of outstanding scientific value or publishable. I believe that provided it is done in a proper scientific manner, any kind of 'research' is worth doing if it gives satisfaction to the investigator and increases his or her understanding and appreciation of a particular field of study such as the Moon.

Lunar topography and mapping

Observing lunar formations and making drawings and charts is a popular form of lunar research and the technique is described in Chapter 5.

Before the advent of the orbiting lunar space craft and close-up photography of the Moon's surface, Earth-based telescopic scrutiny of the Moon was the only means of securing data for constructing maps and charts. Amateurs using telescopes of moderate aperture made real contributions in this field because the giant telescopes at major observatories were rarely, if ever, pointed to the Moon. Nowadays, Earth-based observers using telescopes, however large, cannot hope to see detail as fine as that shown on the splendid close-up photographs of the lunar surface obtained with the space probes. Almost the entire surface of the Moon was covered by the close-up photography, but the area around the south pole and some of the adjacent libratory region of the south-south-east limb was not satisfactorily photographed and is still inadequately charted. This affords a real opportunity for Earth-based observers to still contribute to lunar mapping but the work is bound to be difficult as the area in question lies so close to the Moon's limb.

The Lunar Incognita (south polar) observing and mapping program. The close-up photographs of the Moon secured during the Orbiter IV and V and Apollo missions covered about 99.3 per cent of the lunar surface at least once but the remainder was not adequately photographed. This was because when the Orbiter and Apollo photography was underway this part of the surface was illuminated by very low solar lighting and large areas were obscured by shadows. The Apollo photography did not include the area at all. The region of 'unsatisfactory photography' was therefore 0.7 per cent of the surface, which does not sound like much but represents an area of nearly 105 000 square miles. It includes the immediate vicinity of the south pole and the libratory south-south-east region of the limb beyond the formations Drygalski and Hausen (Fig. 7.1). A small portion of it is never visible from the Earth. The entire region has been called 'Lunar Incognita' and is the last remaining part of the lunar surface for which useful mapping work can be done with Earth-based telescopes.

Because no further close-up photography of the Moon from polar orbits is planned for the foreseeable future, owners of telescopes with apertures of six inches or larger have a wonderful opportunity to cooperate with others in an effort to contribute knowledge of this 'last frontier'. However, it must be emphasised that study of the Lunar Incognita region is difficult, lying as it does close to the limb; surface features are seen in extreme foreshortening and detail is often masked by shadows cast by the abundant lofty elevations there and high elevations will hide what is behind them. The interpretation of what is seen is therefore difficult.

Carefully executed drawings will be of the greatest value because more detail is seen in actual telescopic views under good seeing conditions than in photographs; good photographs are, however, valuable for the positional

Fig. 7.1 The Lunar Incognita region of the Moon (shaded). Shown at maximal favorable libration)

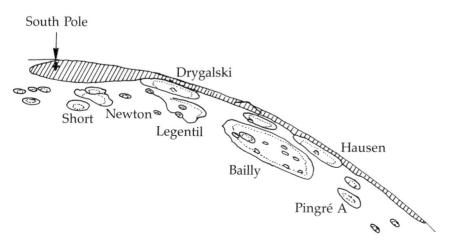

accuracy that they afford. The entire area is strongly subject to librational effects and the best times to observe will be when both libration and solar illumination angles are optimal. These will not, of course, occur in every lunation but the south polar area of the Lunar Incognita can be observed quite well in every lunation. Assuming that seeing conditions are good, the most should be made of every favorable observing time. The examination of stereoscopic pairs of good detailed photographs of the same area taken at different librational states should help in elucidating the precise nature of obscure detail. Also, simulation with models should be useful in the interpretation of what is seen in the telescope. Observations after last quarter when libration is favorable are needed and this requires work in the early mornings in summer and autumn in the northern hemisphere.

Cooperative efforts to map the Lunar Incognita region have been going on for many years on both sides of the Atlantic. Around 1972, the coordinator of the Lunar Section of the British Astronomical Association undertook the construction of a 90-inch polar chart based on observations spread over 20 years with telescopes from 6½ to 12 inches in aperture. The region covered extends in an east–west direction from Demonax–Drygalski down to latitude 75° south. Most detail is in the 80° circle around the south pole and for some distance onto the averted hemisphere as far as was possible.

Also in 1972, the Lunar Recorder of the Association of Lunar and Planetary Observers (ALPO) in the United States of America proposed a cooperative effort to chart Lunar Incognita. Outline forms of the area under different latitude–longitude libration conditions were prepared for distribution to observers and were on a scale of 1 : 2 500 000. Observational details were to be inserted directly on these forms. Visibility ephemerides for Lunar Incognita were published in subsequent issues of the ALPO Journal (*The Strolling Astronomer*). It was pointed out that apart from anything else, the Lunar Incognita region is interesting if only for the fact that here are situated the two highest mountain ranges on the Moon, the Leibnitz and the Dörfel mountains. The Leibnitz range is near the south pole and has never been adequately mapped on a non-foreshortened projection so that its true form can be seen. It is not known whether it is a system of parallel ridges or an irregular upland region. The Dörfels are even less known; they are situated 'beyond' Bailly and since they are invisible at mean libration are not shown on most maps except the special libratory charts of the limb regions constructed by H.P. Wilkins. For all we know, the Dörfels may not even be a mountain range but a profile view of a row of large craters.

At the time of this writing much progress has been made in mapping the Lunar Incognita region and much still remains to be done. Observers with refractors of at least four inches aperture or reflectors of at least six inches aperture can make useful contributions to the program, although again it must be stressed that the work is difficult and observers should be well practiced in observing and drawing 'easier' regions of the Moon's surface. Also, patience is necessary as optimum conditions of both libration and colongitude are infrequent.

Observers wishing to participate in the Lunar Incognita project may wish to avail themselves of the 'Lunar Incognita Observer's Kit' from the Lunar

Recorder of the ALPO. The kit contains a brief set of instructions, a set of 34 outline charts for making drawings (each chart is of a particular area at a specific longitude-latitude libration combination), a graph to help select the appropriate outline chart and a map of the region as seen under optimal libration conditions. The price is $1.00.

The following data should be included with every observational drawing: observer's name; date and time (UT) of observation; type of telescope used, aperture and magnification; quality of seeing and transparency; colongitude and librational state.

The Lunar Incognita observer should consult the February 1972 issue of *The Strolling Astronomer* (Vol. 23, Nos. 7–8) and subsequent issues, many of which contain reports of progress made in the project. The British Astronomical Association's *Guide for Observers of the Moon* should also be consulted. For information on the current status of the program, inquiry should be made to the Lunar Recorders of the British Astronomical Association and the Association of Lunar and Planetary Observers; observations and drawings should be sent to these individuals. The addresses of these organisations will be found in Appendix 7.

Other topographical research. Apart from the Lunar Incognita program, amateur lunar cartography based on Earth-bound telescopic observation can no longer be expected to make major contributions to knowledge of the Moon's surface topography. However, this does not mean that all telescopic observation made from Earth is obsolete; all that has happened is that the emphasis of amateur research has shifted to other areas that are capable of yielding scientific data of permanent value. Before describing these areas, let us pause and consider for a moment the value that telescopic lunar cartography still has apart from the Lunar Incognita project. For one thing, the delineation of lunar formations is a pleasant and absorbing pastime to many amateurs; a well-executed accurate drawing of a surface formation has a value of its own as representing a particular lunar feature under a specific set of lighting and seeing conditions as seen with a telescope of a given aperture.

The systematic sketching of lunar formations is the best way for the beginner to learn his way around the Moon and is excellent observational training. Patrick Moore recommends that beginners should secure a good chart of the Moon (see Appendix 5) upon which about 200 formations are named and also an observing notebook with blank pages, one for each of the named formations. Every one of these should be observed with the telescope and at least two drawings made under different illumination conditions. The drawings need not be artistic masterpieces but, of course, should be accurate and faithful representations of what is seen. This project could take as long as one or two years but it is well worth the effort and is the best way of becoming really familiar with the Moon's surface features, an absolutely indispensable prerequisite if serious research is to be undertaken.

Apart from the purely personal pleasure and the observational practice that lunar drawing affords, there is still a possibility that telescopic observation may yield surprises and new topics for investigation in the post-Lunar Orbiter era. For example, many years ago I wondered why some lunar objects have been studied much more than others. The only representations of many lunar features are the small-scale and often sketchy delineations of them on the various lunar maps; no detailed studies or drawings of many features seem to be available. I therefore compiled a card catalogue of as many published drawings of lunar features that I could lay my hands on. These formations were marked on an outline map of the Moon and shaded lightly or darkly to indicate whether a given feature had been studied intensively or only infrequently (Fig. 7.2). Formations for which I could find no published drawings were not plotted on the chart. Assuming that my survey is a representative random sample, an interesting fact emerges. As can be verified from any chart or photograph, the south-south-west sector of the Moon's fourth quadrant is unique in that it is the only

Fig. 7.2 Distribution of most observed formations on the Moon (shaded)

S

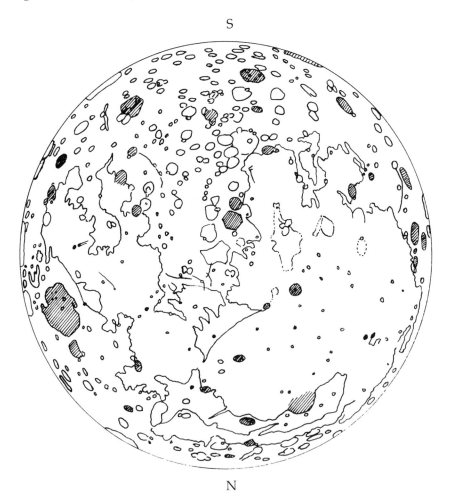

N

one of the eight 45° sectors into which the Earthward hemisphere of the Moon can be divided (by the four diameters passing through the N–S, E–W, SW–NE and SE–NW points on the limb) that is virtually devoid of the dark lunabase (marial) type of terrain. It is a seeming chaotic jumble of crowded ring plains of all sizes. This same sector is also unique in that I found it to consist almost entirely of poorly studied formations. I do not think that this is a coincidence.

Many of the most intensively studied lunar formations are isolated and frequently lie on or near maria and are therefore eye-catching. Examples are Plato, Copernicus, Eratosthenes, Gassendi and Archimedes. (On the averted hemisphere, who has not heard of Tsiolkovsky and Mare Muscoviense, both of which are eye-catching objects? Yet there are far larger and grander formations, almost lost in the general jumble of the averted hemisphere that are hardly known to the average lunar student.) One cannot help wondering whether grandly isolated objects would have had so much attention lavished on them had they been buried amidst similar formations in the crowded fourth quadrant and therefore relatively inconspicuous.

Because of their prominence, even a beginning lunar observer soon gets to know the names of, and easily remembers, the positions of formations like Plato and Copernicus but I wonder if among even fairly knowledgeable lunar observers there are a few who know the exact locations of features such as Abulfeda, Barocius, Clairaut, Heraclitus, Hommel, Licetus, Metius, Mutus, Pontecoulant, Reichenbach or Zagut? All of these and many others are comparable in size and grandeur to Eratosthenes and Copernicus but lie buried among their neighbours in the south-south-west sector of the fourth quadrant. If any one of these had instead been situated right in the middle of Mare Imbrium, I do not doubt that it would have been a well-known and much-studied formation.

A notable exception among the generally neglected formations in the south-south-west sector is the walled plain Stöfler, of which a painstaking study was carried out by Dr H.P. Wilkins, former director of the Lunar Section of the British Astronomical Association, in collaboration with J. Cooke, also of the British Astronomical Association. Even though he published a detailed chart summarising 15 years of work on this formation (see *Journal of the British Astronomical Association*, Vol. 52, p. 64, 1942), Wilkins did not consider it to be exhaustive. In the accompanying text he refers to variable features on the crater's floor, which is traversed by streaks from Tycho and was likened by P. Fauth to the floor of Archimedes. There are dark variable spots that were compared by Goodacre to those in Alphonsus and there are light spots that seem to 'come and go' in a mysterious fashion. Surely, valuable work can be done in confirming Wilkins's observations in this formation alone – maybe some things have changed since Wilkins finished his work.

Well-known formations such as Plato, Aristarchus and Gassendi have been so thoroughly studied and mapped that they must surely be regarded as almost 'mined out.' Almost certainly, little remains to be discovered about them. On the other hand, the south-south-west sector of the Moon's

Earthward hemisphere offers a rich field for exploration by owners of moderate telescopes and may result in discovery of hitherto undetected and interesting phenomena such as 'variability' in the tint of localised areas or 'activity'. Who can tell what may be awaiting discovery or what mysteries lie hidden in the depths of some of the myriad little-known craters and ring structures in the south-south-west sector of the Moon's fourth quadrant?

The accuracy of lunar maps and drawings

Another interesting field of topographical research is comparison of the accuracy of delineation of lunar surface features as seen in the maps and drawings published by different observers. Observers often take considerable pains to delineate lunar detail accurately, yet oddly this is often inserted on a background of coarse detail that is frequently carelessly and inaccurately drawn. Careful comparison of the representations of various lunar features in different maps and charts often reveals astonishing lack of agreement among themselves and inaccuracy in the mapping of even quite large formations on the lunar surface.

Several years ago, Henry Brinton, writing in the British Astronomical Association's periodical *The Moon* (Vol. 9 (1), p. 6, 1960), pointed out the lack of agreement in the various lunar maps in the charting of Nasmyth–Phocylides–Wargentin area of the Moon. Compared with a photograph of the region, the maps of Neison, Elger, Goodacre and Wilkins all differed from the photograph and among themselves with regard to the relative sizes and positions of these three objects and the distances between them. Since this area of the Moon lies close to the limb, the perspective is foreshortened and is strongly subject to librational effects, so that perhaps there is some excuse for these discrepancies in the maps. After all, the authors had undertaken the task of mapping the entire lunar surface as seen from the Earth and were not especially concentrating on this single area. One might imagine, therefore, that large formations close to the apparent center of the Moon's Earthward hemisphere would be the most accurately charted because there is no foreshortening and the effect of libration on the apparent topography is minimal. Let us see.

One of the largest formations on the Moon's Earthward hemisphere is the great walled plain Ptolemaus (diameter over 90 miles), located near the center of the apparent disc. Photographs of Ptolemaus show that under late morning and early evening illumination it is an almost perfect regular hexagon in shape with well-defined vertices. Yet when we consult different lunar maps and compare the delineations of Ptolemaus, considerable deviations from the hexagonal shape are seen in most of them as compared with the photograph and much disagreement among themselves, as a glance at Fig. 7.3 will show. (The gross outline of Ptolemaus has been emphasised in each of the separate sketches rather than the intricate detail of its complex walls.)

Several other examples may be quoted. On two consecutive evenings in November, 1978, I used my eight-inch reflector to observe and make drawings of the crater Godin (diameter 27 miles) under morning illumination. The shape of the formation is an irregular pentagon. On comparing my delineation of Godin with those in the various maps, I found discrepancies. Elger shows it as roughly elliptical, the north–south axis being longer, although in his book he describes it as square-shaped with rounded corners! Neison shows it as diamond-shaped, which is nearer to the true shape, Goodacre as a somewhat distorted circle and Wilkins as perfectly circular. Lohrmann shows one-half of the wall as consisting of three straight sections, the remainder as a semicircle.

Two other formations that I have observed carefully are the craters Pytheas (diameter 12 miles) in the south part of the Mare Imbrium and Calippus (diameter 19 miles) located in the Caucasus mountains. I find that Pytheas is decidedly rhomboidal in shape in agreement with Wilkins and Moore's description (*The Moon*, 1955), although in his map he shows it as elliptical with the north–south axis the longer. R. A. McIntosh also depicts it as

Fig. 7.3 Outline of the lunar walled plain Ptolemaus according to various authors

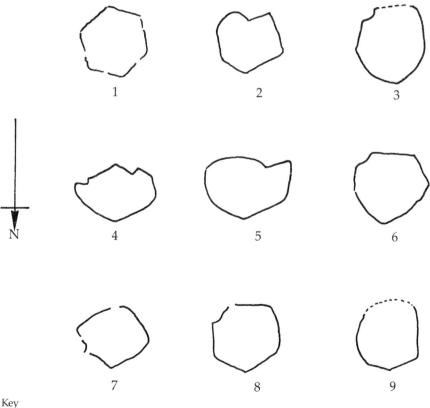

Key

1. From a photograph
2. Beer and Mädler, 1834
3. Lohrmann, 1836
4. Mädler, 1837
5. Neison, 1876
6. Schmidt, 1878
7. Elger, 1895
8. Goodacre, 1931
9. Wilkins, 1958

roughly elliptical in three observational drawings (BAA *The Moon*, Vol. 9 (1), p. 17, 1960).

Calippus is quite obviously irregular in shape. Wilkins describes it as 'somewhat deformed being decidedly oval from east to west' and shows it as elliptical in his map. In an observational drawing, R.A. McIntosh gives a better representation of the shape but still not absolutely correct (BAA *The Moon*, Vol. 9 (2), p. 28, 1961).

One could continue almost indefinitely on this subject. Why do lunar cartographers make such mistakes in delineating the shapes of sizeable lunar formations? Since we cannot claim the effects of foreshortening and libration as causative of the discrepancies in the various representations of Ptolemaus, what other reasons can there be? It would be foolish to suggest that simple error is the cause in the case of selenographers of the stature and reputation of Mädler especially; but what is one to think? It seems unlikely that the use of photographs or 'modern technics' can entirely account for the greater reliability of modern maps such as that produced by the United States Air Force because Wilkins used photographs and worked in fairly recent times like many others. Perhaps preconceived notions of how lunar craters were formed may unconsciously affect the way in which some observers draw crater shapes; an impact enthusiast may be inclined to draw an awkwardly shaped irregular crater with a more 'believable' circular outline.

This is not the place to attempt a complete analysis or explanation of the reasons for errors in the work of lunar cartographers, for the simple reason that in my present state of knowledge of the possible psychological or even physiological factors involved, it would be imprudent to speculate further. The 'psychology of observation' is certainly an interesting field.

The subjective nature of observation

In the previous chapter there were described instances of apparent appearances and disappearances of certain small craters on the Moon's surface. The general conclusion, however, is that no real physical changes have occurred. This is largely because we are forced to depend on the reliability of the selenographers of the past and their delineations, drawings and maps. As is well known, even the best observers and cartographers cannot be relied upon to be absolutely without error. In these cases we are speaking about conclusions based on comparisons of maps and drawings made at different times and often with different types of telescopes with various apertures. Even if the same lunar feature is observed and drawn simultaneously by two observers using similar telescopes and powers, discrepancies may still be noticed between the delineations of the formation by the two observers. This was demonstrated by Arthur Mee many years ago when he and A. Southgate sketched the lunar crater Clavius on the same night but in different localities. There are many craterlets on the floor of Clavius and a comparison of the two drawings shows that each observer had recorded craterlets that the other had omitted and *vice versa*. This plainly shows the folly of jumping to conclusions about disappearances or appearances of small craters based on comparisons of lunar drawings and maps made by different observers of the past. The two drawings of Clavius are reproduced in Mee's *Observational Astronomy* (1893).

Similar discrepancies can be expected in general planetary observation as well as in lunar studies. This has, indeed, been pointed out in the literature more than once. Disagreement is especially prone to occur when detail at the limit of resolution of moderate telescopes is studied. Although published drawings of planets such as Jupiter made by different observers at nearly the same time and with similar telescopes usually show substantial agreement, there are many instances of disagreement regarding the presence, absence or actual appearance of certain features. Different styles of delineation may lead to differences of interpretation. These and many other instances that could be quoted all serve to indicate the essentially subjective nature of lunar and planetary observation.

Clark R. Chapman of the Association of Lunar and Planetary Observers in the United States of America proposed a simultaneous observation program several years ago of certain planets and lunar craters so that discrepancies between drawings of the same object made by different observers at the same time using similar telescopes could be collected, compared and analysed. Thus was a constructive approach initiated to gain some insight into the age-old problem of reliability of lunar and planetary observation. Surely this would be worthwhile repeating with some selected moderate-sized lunar craters? It might also be interesting to construct plaster models of hypothetical lunar features and illuminate them from different angles. The advantage here would be that the topography of the models would be accurately known. To make the study more realistic the models could be observed telescopically from a distance under various 'seeing' conditions. The different appearances presented by the models under various illumination angles could be correlated with the known topography and this might facilitate more accurate interpretation of what is seen in actual telescopic observation of lunar craters.

Observation of central peaks of craters

Knowledge of the numbers and distribution of craters with central peaks is of considerable theoretical importance from the viewpoint of hypotheses of the origin of lunar craters, i.e., whether by meteoritic impact or by endogenous or volcanic forces. Unfortunately, there is much disagreement among selenographers about the number and distribution of craters with central peaks. This is especially so where smaller craters under 15 miles in diameter are concerned.

Another important point is the frequency of central peaks with summit craterlets; better knowledge of the number and distribution of summit craterlets would also have an important bearing on both the meteoritic and volcanic hypotheses of lunar crater formation. Again, there is disagreement as to whether summit craterlets are common or infrequent.

Lunar observers with telescopes of large aperture might be interested in searching for and listing both small craters with central peaks and summit pits on central peaks. Visual observation is especially valuable because even good photographs taken with large telescopes do not show central peaks in small craters that are known to contain them. Visual observations are relatively few in number and some are doubtful.

Examples of craters approaching the 15-mile maximum are Gambart, Helicon, Wöhler, Proclus and Mösting. Observers with large telescopes where there are good seeing conditions could profitably examine craters five miles in diameter and less.

As many small craters as possible should be searched for central peaks when they are about 15° from the terminator, which is about optimal, although this depends on crater depth. Care should be taken as appearances are sometimes deceptive; two close peaks and the space between them may simulate a peak with a summit crater pit.

Many small craters do not have designations and so care must be taken to identify them accurately if central peaks are found. The best way is to draw an accurate chart showing the position of the unnamed craterlet in relation to neighbouring well-known named formations. The usual information recorded in lunar observations such as telescope aperture, magnification, seeing conditions, colongitude and observing location should be recorded. In addition, doubtful or negative results as well as positive sightings should be recorded. When peaks are seen describe whether they are single or multiple, how much of the crater floor is occupied by them, whether they are low hills or prominent peaks. Also, state how positive was the sighting, e.g., certain, moderately certain, etc. An estimate of how much of the crater floor was in shadow should be given.

Because the number and distribution of central peaks and summit peaks are the important factors, any investigation must be fairly comprehensive if it is to have any value.

Domes

The study and charting of lunar domes is a field of research that can be pursued by owners of telescopes of moderate aperture. What exactly is a dome? This question is not trivial because whether or not we record an object seen in the telescope as a dome must depend on how we define it. T. Rackham defines a dome as 'a smooth, circular convex area on the lunar surface, with a height small compared to its diameter'. Essential characteristics are the circular outline and absence of surface irregularities. The disadvantage with this definition is that it is arbitrary insofar as what is seen and classified as a dome in a given telescope is a function of the telescope's aperture; an object seen in a small telescope and classed as a dome may be seen to be not quite circular in outline and to have a rough surface in a larger telescope. Therefore, as seen with the larger telescope, the object does not conform to the defintion of a dome and would not be classified as such. Thus, the bigger the telescope, the fewer are the objects seen with it classifiable as domes.

J. Ashbrook prefers a wider definition that includes irregular outline and a rough surface because, as he points out, an originally 'perfect' dome may have been subsequently altered by igneous activity or it may have collapsed. Some have flattened tops and/or summit pits. If, therefore, we adopt Rackham's narrower definition, certain variant classes of domes may go unstudied. Because the word 'dome' usually refers to a circular smooth regular convex structure, the term 'swelling' might be preferable, as suggested by J.A. Hodgkinson, if we adopt Ashbrook's definition.

The primary criterion of a dome is its very small slope; usually this is not more than 3 or 4° to the horizontal but may be steeper in small parts of it. This is what is responsible for the characteristic pattern of shading at sunrise. Because of their very shallow slopes, domes may be easily 'missed' if looked for under any but very low solar illumination angles. The lunar surface close to the terminator should therefore be searched. Domes on the relatively smooth dark surfaces of the maria are fairly easily seen whereas those on the rougher bright uplands are more elusive.

The beginner at dome study should become familiar with their appearance and visibility by examining some 'classic' domes such as the one in Darwin, which is the largest known, those on the floor of Capuanus, the prominent dome to the east of Milichius, the nearby group of domes north of Hortensius and the well-known example east of Kies. Small hills should not be mistaken for domes. They can be distinguished by their steeper slopes, more angular outlines and less rounded tops. If an object suspected of being a dome casts a pointed shadow, then it can be rejected. Domes are not randomly scattered on the lunar surface but tend to occur in clusters.

Although most of the larger specimens of domes have been recorded, there are many others that need confirmation and there are probably many small ones waiting to be discovered.

Steep slopes on the Moon

Older books on Astronomy often contain artists' impressions of 'lunar landscapes' depicting lunar mountains with very steep slopes and sharply pointed peaks. This impression is a result of the habit of studying lunar surface features under low angle illumination, when long pointed shadows are cast by even trivial elevations. Even a split pea will cast a long pointed shadow if illuminated at a low enough angle. It is now known that lunar slopes are generally quite gentle and that mountains and hills do not have excessively steep slopes or sharply pointed peaks. Steep places on the Moon are quite scarce and few that are visible in telescopes have slope angles greater than 40°, apart from the inner slopes of small craterlets.

The search for steep places on the Moon would seem to be a worthwhile project. The following notes are based on a paper by Joseph Ashbrook read at the eleventh ALPO convention at San Diego, California, USA, on August 22–24, 1963. It was published in *The Strolling Astronomer* (Vol. 17, Nos. 7–8, July–August, 1963). Search should be made for shadows far from the terminator and great care must be exercised in properly identifying features casting shadows. It may be advisable to use some means to cut down glare. Where a shadow is seen this indicates a part of the surface where the slope is greater than the solar illumination angle at that locality. Therefore, the

Sun's altitude gives a lower limit to the slope; conversely, the Sun's altitude sets an upper limit if no shadow is visible. The altitude of the Sun at the given locality must therefore be known and can be calculated from the formula

$$\sin A = \cos B \cdot \sin (L + C)$$

where

A = angular elevation of the Sun

B = selenographic latitude of mountain $(\sin B = \eta)$

L = selenographic longitude of mountain $(\sin L = \xi \sec B)$

C = colongitude of Sun (from the *Astronomical Almanac*)

Occasionally, prominent mountains have steep places on their slopes that still hold shadow for some time after the surroundings are sunlit. Examples are Cape Laplace, Cape Agarum and Boscovitch Beta.

The highest crests of large craters such as Copernicus are among the steepest parts of the Moon. The west wall of Copernicus has sections of crest-line shadow when the Sun's altitude is 42° there. Observation of the decline and disappearance of crest-line shadow in Copernicus on consecutive nights over many lunations would be interesting. Other instances worth studying are Menelaus, Manilius, Agrippa, Pliny and Herschel.

The interior slopes of large craters sometimes have unusually steep places. Good examples are to be found in Langrenus and Eudoxus. Other large craters would be worth searching for lingering small shadows when the Sun's altitude is high.

Search should especially be made for small craters with unusually steep interiors. The best known example is Langrenus M, which still shows shadow on its inner east wall a day or two before full Moon. Schmidt believed that its inner walls had a slope of 60°. This little 10-mile crater is worth further study.

Lunar fault scarps

Every lunar observer knows the 'Straight Wall' on the Moon, the 60-mile cliff running in a roughly north–south direction on the west side of the Mare Nubium between the craters Birt and Thebit. It is really a fault scarp, the lunar surface on the east side being 800 feet lower than that on the west. The 'cliff' therefore casts a strong shadow under morning illumination and the formation is a prominent object in even a small telescope. Under evening illumination, no shadow is cast but the east-facing cliff wall is lit by the Sun and appears as a bright white line against the dark floor of the Mare Nubium.

The Straight Wall appeared to be the only example of a fault scarp on the Moon's surface until H.D. Jamieson found another in the Sinus Roris, which is the area where the eastern end of the Mare Frigoris opens out into the Oceanus Procellarum. The scarp is quite prominent and is oriented in a north–south direction. It was probably 'missed' by other observers because it is visible only under a setting Sun, which in this locality occurs when the Moon is a waning crescent in the sky before dawn.

On the floor of the bright little crater Proclus is what is probably another fault scarp. I was observing the waxing crescent Moon on an evening of good atmospheric transparency and steadiness when the floor of Proclus was about half filled with shadow. Stretching across the floor in a roughly south-east to north-west direction I saw a straight dusky streak which appeared to be a shadow. I saw it again the next night. I have never seen it when the Sun's light is shining from the opposite direction just after full Moon and this is just the behavior that would be expected from a fault. The Straight Wall in Mare Nubium is visible at sunset only because the white 'cliff' face shows up against the dark floor of the mare. Now, the floor of Proclus is white and bright so if there is a miniature 'Straight Wall' on its floor, the illuminated white 'cliff face' would be invisible against the glaring white of the floor of Proclus. The only mention of this feature in the literature is in Goodacre's *The Moon*, where he says that Schmidt's map shows a cleft on the floor of Proclus. I am more inclined to think that it is a fault scarp, for the above-mentioned reasons.

All fault scarps with east-facing cliffs such as the Straight Wall would surely have been noted long ago because of the shadows they cast at sunrise in the first half of the lunation. Any with west-facing cliffs, especially in the eastern half of the disc, might have been missed because their shadows would not be visible until the last quarter phase and later, which would require observing the Moon in the dawn hours. Faults with north- or south-facing cliffs might be very difficult to detect.

Searching for fault scarps would be an interesting piece of research for observers who do not mind staying up into the early morning hours to observe the Moon after last quarter. This would appear to be the time in the lunation when other examples of fault scarps are most likely to be discovered.

The bright ray systems

Research into the Moon's surface features is not entirely a matter of resolving and mapping as much fine detail as possible. The bright ray systems of the Moon, for example, are features of enormous extent, yet, as far as I am aware, there are no really comprehensive or accurate charts of them in existence. The Tychonian system is virtually Moon-wide and the next most extensive system is that of Copernicus. There are many other smaller systems in addition to these. The bright rays are best observed and charted by Earth-based telescopes for the very reason that they are so extensive; if intelligent beings have ever lived on the Moon they would probably have been quite unaware of the existence of the bright rays until they developed means of getting far away from their lunar home and observing its surface from a distance. Large-scale features on the Earth's surface whose existence was previously unsuspected have been discovered only by aerial and space satellite photography. The lunar rays vary in brightness according to the Sun's altitude and it is thought that some rays shift slightly in position during a lunation. Those emanating from the crater Proclus are notable in this regard.

Nobody knows how the bright rays came into existence and many explanations for them have been advanced, none of them entirely satisfactory. Neither is there any satisfactory explanation for the different types of rays; those of Tycho are long and narrow whereas those of Copernicus are less well defined and 'wispy'. Clearly, much work remains to be done. At present there are not sufficient observational data to test the various theories of the origin of bright rays or of their nature. Theoretical models have been put forward suggesting superficial pitting or granulation of the surface to explain their nature, observed features and visibility at different solar illumination angles. Bright rays are most prominent at or around full Moon and are hardly visible when the solar illumination angle is low, although some can be seen when the Sun's elevation is as little as 20°. Curiously, the photometric behavior of bright rays is not always symmetrical with respect to illumination angle, i.e., they are not always equally bright when under similar illumination angle in the lunar morning and evening. Only by amassing sufficient observational and statistical data than are currently available can these and other problems of the bright ray systems be solved.

G.T. Roth suggests that the following program be followed in observing the bright ray systems:

1 The beginning, course and end of each ray should be recorded, also the width and brightness.
2 Look for variations in the brightness of a ray along either its length or width. If so record the position and the time; the latter determines the angle of the Sun's rays.
3 Carefully examine the nature of the lunar surface upon which the ray lies. There may be bright spots and craterlets or slight swellings or uplifts in the terrain that may mimic a ray.
4 In recording the form of the ray state whether it is quite straight or whether it has branches or interruptions; record the positions of these accurately.
5 At times when the ray is invisible look for detail within the ray that shows different colors of the lunar surface.
6 Note the times at which rays appear and disappear in a series of lunations.
7 Color filters can be tried and any changes in a ray's appearance in different colors should be recorded.
8 Studies of the polarisation of light in observations 1–5 may be interesting and worth doing.
9 The selenographical obervations can be usefully compared with similar geological phenomena, if possible.

In addition, statistical study of the distribution of rays may lead to valuable results; for example, their occurrence on low or high ground, on slopes turned towards or away from the origin of the system, or on smooth or rough ground, the presence of discontinuities or curved arcs and the avoidance of surface features. Other things to be looked for are the association of rays with linear arrangements of craterlets, the divergence of rays from points other than the center of the 'parent' crater and their becoming visible only at some distance from the point of origin.

Change and variability on the lunar surface

As shown in the last chapter, the Moon does not seem to be an entirely dead and unchanging world; 'changes' and 'variability' in many localised regions have been reported with different degrees of reliability by both past and present observers.

Transient lunar phenomena (TLP). Of all amateur Earth-based observation of lunar changes, that of transient lunar phenomena offers the best prospect of results having real scientific value. The work requires long-term surveillance of the lunar surface, which for obvious reasons is not practicable from manned orbiting lunar space craft. In any case no further lunar missions are planned for the foreseeable future. Hence, this type of research offers prospects for amateurs to contribute results of permanent value.

The first requirement for successful TLP work is a suitable telescope of large enough aperture. Since many of the TLPs are 'red events', reflectors would seem preferable because the primary image is entirely color-free; the only likely instrumental cause of spurious color is in the eyepiece. Nonetheless, moderately large refractors have been used successfully in TLP studies such as the 10-inch at the Armagh Observatory in Ireland.

The aperture of the telescope must be sufficiently large; although really prominent red TLP such as that seen on the floor of Gassendi in 1966 can be seen directly, the majority require the use of color filters to enhance contrast. Since light is absorbed by filters the telescopic image will be too faint for anything to be distinctly visible if the aperture is inadequate. Patrick Moore says that reflectors of at least eight inches aperture are necessary for reliable sightings although some experienced observers have made definite sightings with six-inch reflectors under excellent seeing conditions. Refractors should have apertures of at least four inches and five or six inches is preferable. Observations should be made under really good seeing conditions and the highest practicable magnification should be used. Red and blue filters are used in searching for red-colored TLP. With the red filter a red TLP would look pale and with a blue filter it would look dark, while the surrounding area would not show any relative difference in brightness with either filter. Thus, if the red and then the blue filter are successively placed just in front of the telescope eyepiece, a 'red event' will be shown up by this differential contrast effect whereas it may have been missed if viewed without the intervention of the filters. However, successful detection of a red TLP in this way depends on the reliability of the observer's visual memory of the appearance of the area being observed with one and then the other filter. This cannot be absolutely relied upon. A better way is to alternate the two filters in continuous rapid succession in front of the eyepiece; owing to the persistence of vision effect a red TLP will show up as a flickering or blinking spot because it appears alternately light and dark with the red and blue filters respectively, while the surrounding terrain will not show this effect. Mechanical devices called 'Moon blinks' have been constructed so as to achieve this rapid alternation of the filters in front of the eyepiece. P.K. Sartory and V.A. Firsoff independently developed Moon blinks. They are not too difficult to construct and have proved their worth in the search for TLP. A Moon blink device is essentially a flat circular drum-shaped chamber containing a rotating disc carrying the red and blue filters. One side of the drum is designed to screw into the telescope draw tube and the other side carries a tube collinear with the draw tube for the eyepiece. The filter disc can be rapidly rotated in either direction to bring one or other filter in front of the eyepiece. This is achieved by a knurled knob on the eyepiece side of the drum that turns a gear wheel inside it which actuates the filter disc (Fig. 7.4). The interested reader will find details about the construction of a Moon blink device in the references listed at the end of this chapter.

Fig. 7.4 Moon blink device

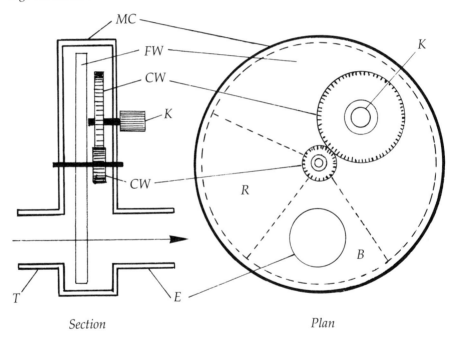

Section   Plan

(arrow shows path of light rays)

R   =Red filter
B   =Blue filter
FW =Filter wheel
CW =Cogged wheels
K   =Knurled head for rotating cogged wheels
E   =Tube for eyepiece
T   =Tube to fit into telescope eyepiece tube
MC =Sheet metal case

In watching for TLP, it is natural to select areas that give the greatest probability of yielding a positive result, such as the 'event-prone' areas of Aristarchus and Gassendi. The trouble with this is that the frequency of TLP in these areas may be falsely exaggerated as compared with other areas because of deliberate selection. Parts of the surface other than these should receive equal time; in their way, negative reports are just as valuable as positive sightings. Certain parts of the Moon's surface show permanent 'blinks' with the red and blue filters and so are not genuine TLP. One example is a streak that crosses the floor of the crater Fracastorius.

In commencing a search, a general survey of known 'event-prone' regions should be done as well as other parts of the Moon's surface. A good plan is to compile a list of features and to become thoroughly familiar with them. Some of these should be 'event-prone' and others not. If a 'blinking' spot is detected or if a colorless mist or obscuration is seen, then adjacent objects should be studied to see if they also show color effects or indistinctness. The effect of changing the eyepiece should also be tried.

A most important matter in TLP research is independent confirmation of a sighting. This means immediately contacting by telephone another observer with adequate telescopic equipment, such as a member of the TLP observing network of the British Astronomical Association. You should mention only the general area where you observed a TLP such as 'floor of Gassendi'. If the second observer records a TLP in exactly the same location as you saw it then this is most valuable confirmation and increases the validity of the sighting immensely. As many as possible of other members should then be alerted by telephone.

The would-be TLP observer is warned that this type of research is both difficult and tedious, necessitating as it does many long hours of what will usually prove to be fruitless search. Also the inexperienced can be misled by spurious color effects, which may be instrumental in origin or due to the Earth's atmosphere especially if the Moon is low in the sky or the atmosphere unclear. It is all too easy to go to the telescope eyepiece expecting to see what you want to see. A false report of a non-existent TLP is detrimental as it will affect the validity of the subsequent statistical analyses of accumulated data from several observers.

Among the TLP the clouds, mists and obscurations that have been reported from time to time are especially elusive. Many years may pass or even the whole lifetime of an individual observer without such phenomena ever being observed by that individual, but you never know your own luck! Any indistinctness or lack of visibility of a formation known to be visible and sharply defined at a particular lunar phase is suspicious. To eliminate possible instrumental effects, the ability to focus sharply on adjacent objects and the effect of changing eyepieces should be tried. Whenever I observe the Moon I always routinely check certain formations, if visible at that phase, for signs of 'activity'. Among these are the south Mare Crisium, the floor of Plato, the interior of Schickard, Herodotus (for the presence of a 'false hill' in its interior) and the region of the 'Cobra-Head', just to the north of Herodotus, for any signs of clouds or mistiness. Other features and phenomena of this sort as described in the last chapter are also worth keeping an eye on and watching for. Perhaps, however, it should be pointed out again that such deliberate selection may give a false impression of the activity of certain areas whereas others may be entirely missed. As with the sighting of 'red events', if anything suggestive of cloud or mist activity is seen, it is most important to secure the confirmation of another observer if possible.

Variable dark and light spots. Several examples of supposedly 'variable' dark spots on the Moon were mentioned in the previous chapter. By 'variable' we mean that the intensity and/or the shape and extent of the spots supposedly varies with time. We must be very careful to distinguish between the regular and predictable changes undergone by these spots during a lunation and irregular or unpredictable changes that do not seem to be correlated with the phase or libration and which in many cases do not seem to be really well established. Much confusion has been introduced into discussions of variable dark spots because of failure to make this distinction. Good examples of both regular and irregular changes in dark spots are the 'pseudo shadows' of Eratosthenes described in the previous chapter.

Of the many bright spots on the Moon, the most intensively studied must be the one upon which the crater Linné stands. Quite apart from the question of whether Linné has undergone a physical change since it was first telescopically observed, the white spot itself is interesting because of the curious changes of size and aspect it undergoes during a lunation. The white Linné spot appears to be largest about four days before and after full Moon and shrinks slightly at full. It is smaller at lunar sunrise and sunset. These apparent changes in size are repeated during each lunation. It supposedly shows dramatic changes in size during a lunar eclipse but this has been disputed. Apparently, it was preoccupation with the Linné craterlet that first revealed the 'variability' of the white spot; this might otherwise have gone unnoticed.

The Linné spot is just one of a host of similar white spots with craterlets on them that are scattered over the Moon's face that undergo similar changes in size during the lunation. Prominent examples are Posidonius gamma, Lassel D and Werner D, all of which are closely similar to Linné. The Mare Serenitatis contains large numbers of smaller white spots. The study of these as a class of lunar objects would be a worthwhile research activity as very little is known about them and their behavior.

Dusky radial bands in craters. Many lunar craters have one or more dusky radial bands on their inner walls, of which some examples seem to be variable. The best known and most observed banded crater is Aristarchus, the inner east wall of which displays several dusky radial bands running up from the floor. Two or three of these are easily visible in a three-inch refractor. The Aristarchus bands appear fairly soon after sunrise, quite suddenly, and are most prominent under a high Sun. Two of the main ones quickly become prominent during a period of one and a quarter hours before the west wall shadow draws back to the long axis of Aristarchus. Some of the bands 'spill over' the wall onto the surrounding terrain.

The inner west wall is poorly displayed owing to the nearness of Aristarchus to the limb. The bands seem to change from day to day in a manner that is not entirely explainable by changing seeing conditions. They seem to have become more prominent since they were first noted. As mentioned in the previous chapter, none of the early selenographers showed them in their charts and maps of the Moon.

Observing with the great Meudon refractor, P.A. Moore saw that the Aristarchus bands had 'fine structure' consisting of small spots, which caused him to suggest that the bands may be composed of chains of dark-haloed craterlets. He also suggested that they may consist of a lowly type of vegetation.

Many other smaller examples of banded craters are known such as Bessarion, Birt, Brayley, Strabo, Isidorus, Isidorus E, Gutenberg A, Thebit A, Lenham, Messier, Pickering and G, a small crater south-west of Langrenus. Larger examples are Menelaus and Hercules.

A.P. Lenham and P.A. Moore recorded several banded craters and there are probably others waiting to be discovered.

A curious property of the dusky radial bands is that in several instances they were difficult to see in large telescopes when first discovered but later were reported as easily seen in smaller telescopes. There is no entirely satisfactory explanation as to what the radial bands are; the objective existence of those in Aristarchus has been questioned. It has even been suggested that perhaps dark crater bands are optical illusions!

A useful line of research would be to confirm the visibility and characteristics of those already known as well as to search for hitherto unrecorded ones.

Physical changes. The only intensively studied and thoroughly documented example of a supposed physical change on the Moon's surface is the alleged disappearance of the crater Linné, described in the previous chapter. However, as was shown, it does not seem as though anything ever really happened there and the apparent disappearance of Linné is based on flimsy evidence.

In order to be quite certain that a physical change of any sort has occurred on the Moon's surface, we must be thoroughly familiar with every single part of the surface under every possible illumination angle and libration state. Clearly, this is virtually impossible for the entire visible lunar surface for a single observer. The best we can do is to become as familiar as possible with a few selected formations and scrutinise them carefully every lunation for years. In so doing, we might increase our chances of detecting and characterising a real change as such but at the same time we would decrease our chances of seeing a change because of restricting our attention to a very small fraction of the surface.

What sort of agencies are likely to cause physical changes? First, it has often been remarked in books about the Moon that the regularly repeated cycle of extreme heat and cold that the surface rocks are subjected to during a lunation must cause cycles of expansion and contraction of the lunar rocks. This would inevitably cause cracking and shattering resulting in landslides, for example. However, for such effects to be unquestionably visible in moderate and even large telescopes, they would have to be on an enormous scale. Nothing like a landslide has ever been detected anyway. In any case they are hardly likely to occur nowadays; the cycles of temperature extremes have been going on for millions of years and surely, by now, all the possible shaking down and shattering of rocks occurred long ago and the Moon's surface has settled down to a state of changeless equilibrium, at least so far as cyclic temperature effects are concerned.

Second, there is meteoric impact. It is quite possible that the Moon's surface is subjected to bombardment by small meteors; since the Moon has no appreciable atmosphere, meteors would reach the lunar surface and not burn up before doing so as they do in the Earth's atmosphere because of frictional effects. It is quite possible that the smallest crater pits shown in the lunar close-up photographs are caused by meteoric impact. Such crater pits are entirely beyond the range of detection of even large Earth-based telescopes and no physical change attributable to meteoric impact has ever been detected with Earth-based telescopes.

Thirdly, there is the possibility of igneous or volcanic activity. There is no compelling reason to believe that lunar vulcanism is entirely extinct and 'Moon-quakes' are quite frequent, as revealed by sensitive measuring instruments left on the surface by the lunar landing craft. However, the Moon-quakes are nowhere near as violent as terrestrial Earthquakes. One possible consequence of igneous or Moon-quake activity is that known cracks and clefts in the crust might extend further or new ones may appear. There have been reports of certain surface cracks that appear to have extended since they were first reported many years ago and even new ones are supposed to have appeared. An apparent disappearance is the case of the cleft first reported by Elger in 1891 on the north part of the floor of Vendelinus. It started from a small crater and extended in a northerly direction. Elger saw it in his 8½-inch reflector and its existence was confirmed by W.H. Maw with a six-inch refractor. More recent observers such as H.P. Wilkins and R. Barker have not been able to find it. In the *Journal of the British Interplanetary Society* for November, 1954, Wilkins describes an observation he made of the area with the Mount Wilson 60-inch reflector at 500× under lighting conditions similar to those when Elger saw the cleft. The cleft was not seen but a long ridge of mounds was observed in the same place. In addition, Wilkins saw two hitherto unrecorded clefts, one on either side of the ridge and parallel to it. Wilkins thinks that a change has occurred in this region.

The reliability of such reports is difficult to evaluate. This is because they rely on the visualisation of often very delicate linear features and this depends on the aperture of the telescope used, the magnification, optical quality, seeing conditions, keenness of the observer's sight, illumination and libration conditions. Attempts to compare the observations of selenographers of the past with those of today are therefore bound to be fraught with error and misinterpretation, so that strictly valid comparisons of old and recent observations of surface cracks are hardly possible. As is well known, it is difficult to find two charts by different observers of the Gassendi and Triesnecker clefts that are in perfect agreement.

Another possible effect of lunar vulcanism is that new domes or swellings may be produced by subsurface gas or magma under pressure insufficient to cause actual volcanic eruptions. In fact, it has been suggested that certain domes at least may be relatively recently formed as they are not recorded by earlier observers, but this seems very doubtful. As previously mentioned, merely because features on the Moon's surface that are easily visible today were not mentioned or recorded by the early selenographers is not evidence that they are 'new'! On the whole, watching out for physical changes on the Moon's surface does not seem to promise much in the way of solid results.

Lunar colors

It may come as a surprise to the beginning lunar observer to learn that there is color on the Moon. When we go to a six- or eight-inch telescope with a dark-adapted eye and view the full Moon with a power of 100–200×, we are confronted with a brilliant sunlit landscape; apart from the shining whites, pale yellows and various shades of grey, there seems to be no hint of color. Many of the great lunar observers of the past and present have nevertheless seen and recorded color on the lunar surface, although there is some disagreement as to the actual tints seen in a given area. This is not surprising as the eyes of different people have different sensitivities to different wavelengths of light, and color vision during the lifetime of a given individual may undergo changes. The study of lunar colors is therefore partly subjective but the use of color filters to enhance contrast greatly assists in the observations and largely removes the subjective effects.

One of the most obviously colored regions on the Moon is the roughly diamond-shaped area north-east of the crater Aristarchus. To most observers it appears to be of a yellowish or khaki hue. A similar tint but more inclining towards greenish is almost equally prominent in the rugged area to the south-west of the lunar Apennines where they come close to the Haemus mountains. The best time to see these is when the Moon is high in the sky on a clear night.

H.P. Wilkins reports greenish tints in the Maria Crisium, Humorum and Tranquillitatis and inside the craters Grimaldi and Ptolemaus. Lunar colorations are affected by the angle of illumination; W. Haas summarises their behavior as follows. Colors seen under low angle illumination are short-lived but fairly prominent. These are usually greens, occasionally purples or blues and sometimes browns. Such colors are usually seen on dark-floored craters when near the terminator. Colors viewed under high illumination are difficult to see but last longer. Blues and purples are thus seen in dark areas and greens in extensive dark areas.

Some lunar colors undergo regular changes during the lunation. A well-known example is the floor of Ptolemaus, which changes from grey to olive green during the lunar morning to yellowish as evening approaches. Other colors seem to undergo much longer cycles of change.

Many lunar colors, especially the greens, may be due to color inherent in the lunar rocks. However, V.A. Firsoff states that the strong greenish tints sometimes seen near the terminator rapidly disappear as the Sun rises higher. He notes that snow reflects green light most strongly, unlike white quartz which appears very bright in red and blue. The telescopic view of distant hill sides covered with hoar frost and their appearance in color filters is not unlike that of the crescent Moon and he considers it not impossible that the fugitive green tints near the Moon's terminator may be due to snow or solid carbon dioxide. Most bright crater rims, rays and crater floors are also bright in green.

Even more interesting and worth looking for are colors that may be due to the different degrees of scattering of light of various wavelengths in a rarefied atmosphere or to localised gas emissions. V.A. Firsoff reports an interesting observation of sunset on the lunar Apennines as seen through a red-colored filter. With this, the mountain peaks shone brilliantly but appeared dull when viewed with a violet filter although the Moon's limb and the poles appeared undiminished. This observation indicates a pre-dominance of red in the light reflected from the mountains, which reminds us of the redness of terrestrial sunsets. It therefore could be due to the presence of an atmosphere. The redness of the terminator was noticed by D.P. Avigliano, the American student of lunar color, and is also indicative of a tenuous atmosphere, which would scatter the shorter wavelengths (blue and violet) but would allow the longer wavelengths (red) to pass through.

The yellowish area north-east of Aristarchus, as expected, usually looks dark if viewed with a violet filter. Yet, V.A. Firsoff has seen it looking pale with a violet filter as has also A.G. Smith, who saw it pale using an ultra-violet filter with an electronic image converter. These observations are exp-lainable on the basis of light scattering by temporary gaseous emissions in these areas. Violet light observations of other parts of the Moon's surface likewise indicate gaseous exhalations, which have the effect of scattering light of short wavelengths. One of these is a line of dark marial material, the Mare Veris as it is called, close to the Mare Orientale on the east limb of the Moon. This is a prominent feature when observed directly or with red, green or yellow filters but it is frequently quite pale when viewed with blue or violet filters and occasionally is nearly invisible. Firsoff considers that the best explanation of this phenomenon is obscuration by gas. The indistinctness of clefts on the floor of Alphonsus when photographed in violet light, again indicative of gaseous emission, was mentioned in the previous chapter.

The foregoing examples are only a very few of the many confirmed observations of color on the Moon's surface and should serve to convince anyone of the genuineness of lunar colorations.

We have noted the subjective nature of such research carried out by telescopic observation with the otherwise unaided eye. Lunar colors are faint and usually difficult to see because the overwhelming glare of the telescopic image tends to obscure subtle tints and hues. This may be coun-teracted by stopping down the aperture. However, D.P. Avigliano says that the colors are more difficult to see if reduced aperture is used. He maintains that using a reflector of eight to 10 inches aperture of focal ratio F/8 to intensify colors is necessary. One possible way to get around these contradic-tory requirements is to use the full aperture with medium magnification and to restrict the view to a small portion of the Moon's surface. One way to do this is to move the eye away from the eyepiece. The exit pupil contracts quickly but the image size remains unchanged so that any specific object or area can be isolated from the rest.

The most important aid in lunar color research is the use of properly designed color filters. Filters of the type used in color photography are unsuitable; although they give preference to one color, too wide a range of wavelengths is transmitted. Filters that give good results in lunar color research are those that separate the three color responses of the retinal cone cells of the human eye. These filters allow only a narrow range of wavelengths to pass. For example, if a red filter is used and red is present in the object being studied it will appear bright when compared with the surrounding terrain, but when viewed with green or blue filters, which absorb red wavelengths, the feature will look dull. A blue or violet filter will make a yellow area look dark. There are two kinds of filter that are useful in lunar color work – the so-called monochromatic filters and the tricolor separation filters. The monochromatic filters have a narrow pass band, meaning that they transmit a relatively narrow band of wavelengths and so isolate almost pure spectral colors. The tricolor separation filters, such as the red, green and blue filters comprising the Dufay tricolor filter set, have wider pass bands than the monochromatic filters so that they do not transmit pure spectral colors. There is a slight overlap at the extremes of their transmission ranges but the transmission peaks are well marked. These filters effectively separate out the three main color responses of the human retina. When filters of either type are used, a green-tinted object on the Moon will look darker than its surroundings when viewed through a red filter but will look bright with a green filter. With the tricolor set a yellow area will appear bright with the red and green filters and dark with the blue, while a violet-tinged area will appear dark with both red and green filters. Even white areas that look alike without filters will behave differently when viewed with color filters, such as the different appearances of snow and white quartz in a green filter mentioned earlier. Blacks and greys that look similar may also be differentiated with suitable filters.

In practical work with filters it is a nuisance to be continually changing them by hand. It is a good idea to mount a set of filters in a device like the 'Moon blink' used to detect red TLP so that they can be rapidly and conveniently changed.

Observing lunar eclipses

A lunar eclipse occurs when the Moon passes into the shadow of the Earth cast by the Sun. As explained in Chapter 2, an eclipse of the Moon can occur only at the time of full Moon and when the Moon is at, or very close to, a node so that the Sun, Earth and Moon are nearly in a straight line.

An eclipse of the Moon can be seen from an entire hemisphere of the Earth and will look the same at every station. Hence, at any given place, eclipses of the Moon are decidedly more frequent than eclipses of the Sun, which can only be seen within a relatively narrow band of the Earth's surface. Generally, two full Moons pass through the Earth's shadow each year.

Since the Sun is a much larger body than the Earth the shadow of the Earth will consist of a dark cone-shaped umbra and a lighter penumbra. Sometimes the Moon will enter the penumbra only and so there will be a penumbral eclipse. At such times it is often difficult to believe that an eclipse is in progress as the penumbra is quite pale. Umbral eclipses are more interesting and spectacular and are partial or total depending on whether the Moon partly or completely enters the umbra.

If the Earth had no atmosphere the Moon would be absolutely dark and completely invisible during a total eclipse. The Earth's atmosphere, however, acts as a giant lens and refracts the red wavelengths of the Sun's light towards the Moon during an eclipse with the result that the eclipsed Moon often has a coppery red color.

The brightness of the totally eclipsed Moon depends on the state of the Earth's atmosphere. Sometimes the Moon can look quite dark; the eclipse of 1884 was unusually dark and this was undoubtedly caused by volcanic dust in the upper atmosphere from the eruption of Krakatoa in the previous year. Another dark eclipse occurred in 1950 and this time was due to atmospheric dust originating from widespread forest fires in Canada. The total eclipse of June 25, 1964, was very dark. This was caused by large quantities of dust sent into the atmosphere by a volcanic eruption in the East Indies that had occurred previously and which had the effect of cutting off much of the Sun's light. Patrick Moore reported that he was unable to see the Moon during totality even with his 12½-inch reflector. In more recent years the total lunar eclipse of July 5–6, 1982, was very dark and this was undoubtedly due to large amounts of dust thrown into the atmosphere of the Northern Hemisphere by the volcano El Chichon, in Mexico, which had erupted in March of that year. Six months later, on December 30, 1982, there was another very dark total lunar eclipse which in several ways resembled an exceptionally dark eclipse 19 years previously to the day, also a result of volcanic dust in the Earth's atmosphere. One American observer, John Bortle, 'a veteran of twenty lunar eclipses', wrote that on the Danjon scale (described later in this section) of expressing the brightness of a lunar eclipse, 'only the $L = 0$ description will fit'. From a study of the reports sent to the office of *Sky and Telescope* magazine it was apparent that the eclipse of December 30, 1982, ranked with approximately 12 very dark total lunar eclipses that occurred during the last 400 years.

The totally eclipsed Moon can also appear so bright that it is difficult to convince people that the Moon is eclipsed. Such a bright eclipse occurred in 1848, the Moon appearing a strange blood red color. A very curious reflection phenomenon was observed in 1895 when the outline of the African continent was distinctly seen on the copper-colored ball of the totally eclipsed Moon. Not only red, but also blue tints are sometimes seen as in the total eclipse of January 29, 1953, and the partial eclipse of July 15, 1954.

These phenomena are unpredictable and have nothing to do with the Moon itself; it is the Earth's atmosphere that is responsible. It is quite interesting to attempt to correlate the appearance of total lunar eclipses with the prevailing terrestrial atmospheric conditions.

Interesting though these phenomena are, our main observations during an eclipse should be of the Moon's surface features. We must not forget that those parts of the lunar surface entering the umbra of the Earth's shadow will experience a sudden drop in temperature. For a spot near the lunar equator this can be from a temperature near that of boiling water to minus 100°F in less than an hour. This is because the Moon has no atmospheric 'blanket' to protect it from sudden temperature changes. On emerging from the shadow the temperature of the same area will rise again to its previous value just as rapidly. Unusual effects may be produced by these sudden thermal shocks and might cause changes in the appearance of the supposedly 'variable' dark markings in the craters Endymion and Eratosthenes and the dusky radial bands in Aristarchus. The dark floors of Plato and Grimaldi may also be worth looking at during and after an eclipse. The white nimbus surrounding Linné is said to enlarge during an eclipse although some observers have not noticed any change. It is very easy to be deceived in such matters. A good plan would be to carefully study areas suspected of variability for several weeks before an eclipse is expected and to become thoroughly familiar with their appearance under various normal lighting conditions and then to look carefully for any changes immediately before and after they enter the shadow.

The timing of shadow contacts of selected lunar features can provide data from which the diameter of the Earth's shadow on the Moon can be calculated; the shadow diameter differs slightly from one eclipse to another. At the Moon's average distance from the Earth, the diameter is about 5700 miles. The actual diameter of the Earth's umbral shadow on the Moon is always larger than would be expected from the geometry of the eclipse. Predictions of lunar eclipses make approximate allowances for this. Curiously the degree of enlargement varies from one eclipse to another; the reason for this is not fully understood. The enlargement can be deduced from carefully timed observations of the four principal umbral contacts during an eclipse and from the timing of umbral contacts with craters. The method is to select several small to moderate-sized craters scattered over the Moon's disc. During the partial phases of a total eclipse the times are noted when the edge of the umbra makes contact with the center of each of the selected craters. In the case of the large craters, the times that the edge of the shadow makes contact with one wall and then the opposite wall are noted and averaged and this is taken as the time of contact with the center of the crater. Craters that have been suggested for timing of umbral contacts are: Grimaldi, Aristarchus, Kepler, Copernicus, Pytheas, Timocharis, Tycho, Plato, Aristoteles, Eudoxus, Manilius, Menelaus, Plinius, Taruntius and Proclus.

For observing lunar eclipses it is not necessary to have a large telescopic aperture and low enough powers may be used for the whole disc of the

Moon to be seen in the field of view. The edge of the umbral shadow is ill-defined so that using high powers contributes nothing to the observations.

The following data should be recorded during a lunar eclipse.

With the naked eye, the definition of the umbral edge and the color and density of the shadow during the eclipse should be recorded. The simplest way to rate the darkness of the eclipsed Moon is to make an estimate based on the five-point scale devised by A. Danjon. The darkness of the eclipse is expressed by an L number:

L = 0. Very dark eclipse. The Moon almost invisible particularly at mid-totality.

L = 1. Dark eclipse. The color is usually grey or brown and the surface details are difficult to see.

L = 2. Deep red or rust-colored eclipse. There is a very dark center to the shadow while the edge of the umbra appears relatively bright.

L = 3. Brick red eclipse. There is usually a bright or yellow rim to the shadow.

L = 4. Very bright copper-colored or orange eclipse. The shadow rim is bluish and very bright.

With the telescope the following should be recorded:

i Contact times of the umbra with the lunar disc.

ii Crater contact times of the umbra.

iii Definition of the edge of the umbra; this is variable and unpredictable. Note any irregularities in its outline and the width of that part of the edge that shows any gradation of density.

iv Whether the shadow edge is uniformly colored and whether it is the same as the inner part of the umbra.

v The visibility of selected bright and dark areas during their passage through the shadow. Areas on the limb and well away from the limb should be selected. Notice any unusual appearances in other parts of the disc.

vi Any difference of density or color between the first and second half of the eclipse.

Also, the aperture and magnification(s) of the telescope should be included as well as atmospheric conditions and whether there was a lunar halo, and the latitude, longitude and height above mean sea level of the observer. Timings should be in UT and given to the nearest 0.1 minute.

Occultations of stars

Since the Moon is the nearest celestial body to the Earth and takes about a month to make one complete orbit around the parent body, it is easy to see its 'drift' in an easterly direction against the background of the fixed stars. If its position with respect to a certain fixed star is noted on a given evening, then on the next evening at the same hour the Moon will be seen to have moved about 12° east of its previous position.

The Moon's apparent path in the sky lies within the Zodiac, which is a belt of sky about 18° wide. The planets of the Solar System also move within the Zodiacal band. Occasionally, because of the inclination of their orbits, Mercury and Pluto will be found outside the limit. There are 12 constellations of the Zodiac – Aries, Taurus, Gemini, Cancer, Leo, Virgo, Libra, Scorpio, Sagittarius, Capricorn, Aquarius and Pisces. Some of these contain quite bright stars and, every now and then, the Moon will pass in front of one or other of these during its eastward drift in the sky. Thus, the star will be temporarily eclipsed or occulted to use the more usual term. Occultations of stars can be predicted in advance and these predictions are published annually in the handbooks of astronomical societies such as the *Handbook of the British Astronomical Association*. Since the positions of the fixed stars are known very precisely, the accurate observation and timing of occultations gives valuable information about the exact position of the Moon's limb and hence of the Moon itself, which is not known to the same accuracy as the fixed stars. Data of this sort are of value to the professional computers who calculate the dates and times of occultations in advance. It is a pity that not many amateurs observe and time occultations on a regular basis because such observations are of real scientific value. All the equipment required is a small telescope and a good stop-watch. Because of the Moon's eastward drift in the sky it follows that when a star is about to be occulted it will disappear at the Moon's east limb. This is called immersion. Likewise, it reappears at the west limb and this is called emersion. The Moon appears to move easterly at such a rate that it covers a distance among the stars equal to its apparent diameter in about one hour. Hence, the longest time that a star can be occulted is about an hour. Before full Moon, a star will be occulted at the dark east limb of the Moon, or if after full the limb will be bright. The opposite applies for emersion.

Since a star appears as a bright point of light without appreciable angular size, its disappearance when occulted by the Moon's dark limb before full is startlingly sudden; at one instant it is there and at the next it is gone. Its reappearance at the other limb is equally sudden. Incidentally, this suddenness of a star's disappearance and reappearance is evidence of the virtual complete absence of a lunar atmosphere. If there was an appreciable lunar atmosphere these phenomena would be more gradual owing to atmospheric refraction of the star's light. The star's light would grow dimmer and would redden instead of suddenly snapping out when the Moon's limb occults it.

Occultations are usually predicted in observers' handbooks to a 10th of a minute so that in preparing to observe an occultation it should be sufficient to be at the telescope about 10 minutes beforehand. Immersions are easier to observe and time accurately because the star can be seen gradually approaching the lunar limb, but emersions are more difficult because the sudden reappearance of the star from behind the Moon comes as a surprise and there may therefore be some slight delay in timing the event.

In timing an occultation, the stop-watch should be carefully wound up and started from some kind of accurate time signal. A small telescope and a low power are quite sufficient for observing and timing occultations.

271

Another important prerequisite if occultation timing is to have scientific value is for the observer to know his exact location in latitude and longitude and the height above sea level. The geographical position should be determined to within approximately one second of arc in geodetic latitude and longitude, and the height above sea level determined to within 50 feet. This is necessary for accurate reduction of the observations to be carried out. Your position may be found by consulting large-scale survey maps of your area, which should be available from a reference library or from the surveyors' departments of your local council. In Canada, such maps are prepared by the Department of Mines and Resources and in the United States, the Department of the Interior publishes topographical maps. The accurate determination of position is important because, owing to the parallax effect, observers in different parts of the country will have slightly different views of an occultation.

Your latitude makes a difference because of the parallax effect; an observer in, say, New York City may see the Moon pass north of a star whereas to an observer in Ottawa, the star could be occulted.

Your longitude affects the time at which an occultation occurs. The further east you are the later is the occultation.

Sometimes a star will just touch or 'graze' the Moon's north or south limb and this is called a grazing occultation. Grazing occultations are especially interesting and valuable because they may enable minor errors to be detected in the calculations on which predictions are based and hence the Moon's position may be more accurately determined. As an illustration of this may be mentioned an occultation that occurred on February 20, 1964, that was observed from England. The prediction was that the star would pass behind the Moon's south pole but the group of observers who were watching could see that it would actually be a grazing occultation. The occultation limit, i.e., the line between the star just being occulted and a 'near miss', extended across the south of England. One of the observers watched from right on the limit while a group of observers watched from a place several hundred yards inside this limit. The party observed a graze while the single observer on the limit saw a 'miss'. Therefore a slight error had been made in the prediction calculations and so the position of the Moon could be corrected.

In recording an occultation the timing should be within a fraction of a second. Other data that should be included are the name or number of the star, the observer's latitude and longitude and the height above the mean sea level of the observing site, the seeing conditions and any unusual appearances. Sometimes a star may be momentarily hidden by a mountain on the Moon's limb to reappear again for an instant before finally disappearing, so that it seems to blink. Such a phenomenon is unpredictable because the exact position of the Moon and the profile of the limb cannot be known with absolute precision.

Lunar photography

Photography of the Moon through the telescope is an interesting and challenging activity and good photographs can be obtained with simple cameras used in conjunction with good telescopes. Many observers have found Moon photography to be an interesting hobby and there must certainly be a sense of accomplishment attending production of good pictures of the Moon's surface features. It goes without saying that however good the quality of Earth-based lunar photographs, they cannot hope to compete with the close-up photographs taken with the Lunar Orbiter space craft so far as revealing fine detail is concerned. However, good amateur photographs provide material from which accurate outlines of lunar formations can be traced and positional accuracy assured.

I do not myself have any experience in the field of lunar photography so am not qualified to write on the subject but details and procedures will be found in the article on lunar photography by Henry Hatfield in the British Astronomical Association's *Guide for Observers of the Moon* and in the two articles on lunar photography mentioned in the reading list following this chapter. However, the following general notes will give the reader an idea of the principles involved.

The simplest type of lunar photography can be done with any ordinary camera without modifying it. First, focus the camera for infinity. Now point the telescope to the Moon and, looking through the eyepiece, get the image as sharp as possible. A normal-sighted person will thus focus the telescope for infinity. Take the camera and hold it up to the eyepiece, positioning it carefully. Take a snapshot exposure. It is difficult to estimate what exposure you will need so it is best to try several and this should result in at least one correctly exposed negative. This simple procedure will not yield masterpieces of lunar photography but will give valuable practical experience for a beginner.

To avoid possible disappointment, the would-be lunar photographer should realise that no photographic emulsion yet devised records the finest detail resolved by a given telescope that can be seen by visual observation, with the same telescope, during those moments of perfectly steady seeing that somtimes come on a good night. There are two principle methods used in serious lunar photography:

1. *Prime focus photography.* As the name suggests, it is the primary focal image of the Moon produced by the telescope objective that is photographed so that the camera lens need not be used at all. If you have a 35 mm single lens reflex (SLR) camera to couple to your telescope you will be well equipped to take prime focus photographs of the Moon. The telescope objective itself acts as a giant telephoto lens. Remove the telescope eyepiece and the lens from the camera. Attach the camera to the eyepiece tube of the telescope by one of any of the camera-telescope adapters that are commercially available and point the telescope to the Moon. Focus the Moon's image on the focusing screen of the camera, using a magnifier to ensure that the image is accurately focused, and make an exposure. You know the F-ratio of the telescope objective so you should give the same exposure as you would give any other body illuminated by bright sunlight at that focal ratio. Avoid shutter vibration by using a cable release. With ordinary telescopes having focal lengths around 45–60 inches, this will give an image of the Moon on the film about half an inch in diameter or slightly more. This will show

crater detail quite sharply. The roughly half-inch negative image of the Moon can be easily enlarged up to six or eight inches in diameter if care is taken with focusing the image prior to making the exposure and avoiding camera vibration when releasing the shutter.

2. **Projection methods.** Larger images and more detail can be recorded by using the so-called projection method of photography. A larger image of the Moon will be projected if the eyepiece of the telescope is left in position. The projected image can be caught on the film just as in the prime focus method. Some other means will probably have to be found to attach the camera to the telescope as the eyepiece occupies the tube that the camera adapter slides into.

The eyepiece must be of the highest quality, such as an orthoscopic, for best results. Achromatic Barlow lenses or special projection lenses can also be used with ratios of 4:1 and 6:1. Using, say, a 4:1 projection lens attached to a three-inch refractor of focal length 45 inches (F/15), the effective focal ratio will be increased to F/60 and the image of the moon will be approximately two inches in diameter. The image, will of course, be much less bright than the prime focus image; in this case it will be only one-sixteenth as bright as the one-half inch image formed at the prime focus of the three-inch objective, owing to the inverse square law. This means that longer photographic exposures will be needed and this may result in blurring of the image due to the apparent motion of the Moon due to the Earth's rotation if a stationary telescope is used. However, if the exposure can be kept to about one-fiftieth of a second, depending on the lunar phase and film speed, then a stationary telescope can be safely used. The detrimental effects of atmospheric turbulence on the image are also greatly increased by longer exposures. Several exposures should be made. Hopefully, one of these will not be too badly blurred by atmospheric unsteadiness. Fast films can be used to keep exposure times short in the projection method and the generally coarser grain of such films need not be detrimental to fine detail in the image because of the larger image size as compared with prime focus photography.

If exposures of several seconds need to be made then the telescope must be driven to counteract the motion of the Earth; better still, for best results, it should be adjusted to counteract the apparent slow eastward drift of the Moon due to its own orbital motion.

The calculation of the correct exposure time can be tricky and depends on the effective focal length of the optical system, the phase of the Moon, the height of the Moon above the horizon and the speed of the photographic emulsion used. The reader is referred to the works dealing with lunar photography at the end of this chapter for details.

Films, development and enlarging. Using the right film is an important factor in successful lunar photography as it is in any branch of photography. A compromise between opposing requirements has to be reached: fine grained films make for capturing fine detail on the resulting negative but the generally slow speed of such films will make relatively long exposures necessary. This increases the risk of blurring the negative detail, principally because of atmospheric turbulence. Errors in telescope tracking or accidental vibration are also more likely to harm the image. Fast films will cut down on exposure times but the relatively coarse grain may not permit rendering of the finest detail, especially in prime focus photography. Then there is the question of the contrast characteristics of the film and the brightness range of the Moon's surface. At full Moon, when there are no shadows, the surface is equally illuminated and the brightness range is probably no more than about 50 to 1. This is due solely to local differences of reflectivity or albedo of the surface. At other phases, especially at and near the quarters, the range may be about 1000 to 1 as a rough estimate. This is because at the terminator the surface is obliquely and therefore dimly illuminated. No photographic emulsion is capable of recording such an enormous brightness range while maintaining good resolving power. A single uniformly exposed negative will therefore not give a correct representation of the Moon's surface.

One way out of this difficulty is to use a camera with a specially designed focal plane shutter that gives differential exposure of different parts of the Moon's disc if the quarter phases are being photographed; the dimly lit surface near the terminator is given a longer exposure time than the brighter limb regions.

Fast emulsions have a considerable latitude in exposure time and give negative image contrast that is less than that in the object itself. Hence, it is possible to obtain negatives registering the entire brightness range with ordinary uniform exposures. However, when it comes to making a print from such negatives, graded exposures of different parts of the image onto the paper are necessary. This may be done by using specially cut cardboard masks or even one's hand in the method of graded exposure during printing called 'dodging', a trick known to every practical photographer. Suitable films and speeds for the beginning lunar photographer would be 17 DIN (50 ASA) panchromatic cut films or plates, which are easily obtainable. When developed with fine grain developers they are least likely to give problems. Their exposure time latitude is good.

For those who wish to use the shortest exposure time, emulsions of 23 or 25 DIN (about 200 ASA) may be tried. This has coarse grain but will not adversely affect the final results. Also worth trying are fine-grained contrasty films with thin emulsions of speed 12 to 15 DIN. The speed can be increased by using certain developers like Neofin.

The above notes are intended only to be a rough guide and nothing can replace actual trial and experimentation. Only in this way can the lunar photographer determine which films and developers best suit his needs.

In summary, the two principal methods of lunar photography with their advantages and disadvantages can be compared.

In *prime focus photography* the focal distance is equal to that of the telescope objective and will be from about 40 inches and up. This method yields smallish but bright images with good definition so that short exposure times can be used, often less than one second, so that a stationary telescope can be employed. Film emulsions of speed 13–17 DIN are therefore suitable and enlargements of from 10 to 20 times the negative image size may be made. The effects of atmospheric turbulence on the negative image are noticeable in spite of the short exposure times.

In *projection photography* the effective focal length will be much greater than that of the telescope objective lens because of the projection lens or lenses used. The image is much larger and definition good. Exposure times may be from one to three seconds so that a guided telescope should be used, particularly with the longer exposure times. Film emulsions of speed 17–21 DIN are suitable. Enlargements of from five to 10 times the negative image diameter may be made. The effect of atmospheric turbulence on the image is considerable.

Determination of the heights of lunar surface features

The determination of the heights of lunar mountains and other surface features is a field of research capable of yielding valuable scientific data; there must be hundreds of lunar features that have not yet had their heights measured. This type of work will appeal to the observer who likes mathematics. There are two steps in the determination of the height of a lunar feature:

    1 Measurement of the length of the shadow it casts.

    2 Calculation of the height of the feature from its shadow length.

Measuring the shadow length. There are four methods of measuring the length of the shadow cast by a lunar elevation. In increasing order of accuracy they are as follows:

i Estimating by eye. This is rather a crude method but it is suitable. One merely compares the shadow length of the feature with an object of known dimensions, such as the diameter of a nearby crater. Needless to say, the crater selected should not be seen in foreshortened perspective. If the shadow length is estimated to be one-half of the diameter of, say, a 20-mile crater then the shadow is 10 miles long. Improved accuracy is afforded by making several estimates of the shadow length by comparing it with other nearby craters of known diameter. A mean value can then be calculated. The highest practicable telescopic magnification should be used that seeing conditions will allow.

ii Photographic methods. This is similar in principle to the eye estimating method but it yields much more accurate results. For best results the scale of the photograph and the time when it was taken must both be accurately known. Very small photographs are unsuitable as are greatly enlarged prints in which grain is noticeable and shadow edges ill-defined. There is an optimum scale between these extremes. Only the best obtainable photographs should be used. The diameter of a nearby crater, which is accurately known, is measured at the photograph with a transparent millimeter scale. Then the length of the shadow of the feature whose height is desired is measured. It is then a simple matter to calculate the length of the shadow from the formula:

$$\text{shadow length} = \frac{C \times a}{b}$$

where

$\quad\quad\quad C$ = crater diameter in miles
$\quad\quad\quad a$ = length of the shadow in millimeters
$\quad\quad\quad b$ = diameter of the crater in millimeters

iii The drift method. To measure shadow lengths by this method, the telescope must be equatorially mounted, motor driven and the alignment must be stable. The highest possible magnification should be used that prevailing seeing conditions permit. The eyepiece of the telescope is provided with a fixed north–south thread at the focus. A stop-watch is also needed.

The Moon's own orbital motion causes it to drift in a west to east direction in the telescope field when the drive is *on*. The rate of drift is something less than one kilometer (5/8 mile) per second. Therefore, with the telescope drive on, first find how long it would take for one lunar radius to drift across the eyepiece thread. This is done by using the stop-watch to find how long it takes for two reference points to drift across the thread, taking the average of several measurements. By taking the ξ coordinates of the reference points the time taken for one lunar radius to drift across the thread can be determined.

The time taken for the tip of a shadow and its base to drift across the thread is then measured. The mean of several timings should be taken. The drift time can then be converted into the shadow length in terms of the Moon's radius.

iv Filar micrometric methods. The filar micrometer is a measuring device for determining the angular dimensions of an object viewed in the telescope or the angular distance between two points, such as a double star. The device attaches to the eye-end of the telescope and the eyepiece is inserted. Upon looking through the eyepiece, three fine straight threads will be seen in the field of view. Two of these are fixed and perpendicular to each other and usually intersect at the center of the field. The third, parallel to the vertical thread, is moveable to the right or left by means of a very accurately constructed micrometer screw while remaining exactly parallel to the fixed vertical thread (Fig. 7.5). The screw threads measure 0,5 millimeter each and the total amount of horizontal travel is 24 millimeters. To make measurements of shadow lengths the shadow is positioned with, say, the base touching the fixed vertical thread and the moveable thread is adjusted by the screw mechanism until it just touches the tip of the shadow. The thread value is noted, i.e, the reading on the micrometer scale. A similar measurement of a crater of known diameter is made and the thread value for this noted. The length of the shadow is given by the formula:

$$\text{shadow length} = \frac{C \times a}{b}$$

where

C = crater diameter in miles
a = thread reading for the shadow
b = thread reading for the crater

The telescope must, of course, be on a motor-driven equatorial mount when micrometer measurements are made.

Calculation of the height of the lunar feature. First, the position of the feature whose height is desired must be accurately determined, i.e., the selenographic latitude and longitude. These are obtained from the feature's ξ and η coordinates, which are best taken from D.W.G. Arthur's *Orthographic Lunar Atlas* or the IAU Atlas of 1935 but the latter is not so reliable in the limb regions. The feature's selenographic latitude (L) and latitude (B) are calculated from the following formulas:

$\sin L = \xi \sec B$
$\sin B = \eta$

Next, take the values of the following variables from the *Nautical Almanac* and *American Ephemeris*:

L' = Earth's selenographic longitude
B' = Earth's selenographic latitude
C = Sun's selenographic colongitude
B'' = Sun's selenographic latitude

The interpolation is made easier if the Universal Time (UT) is expressed as a decimal fraction of a day, for example June 12 $6^h\ 0^m$ is equal to June 12.25 (see conversion table in the *American Ephemeris* for 1962, p. 456). The angular

Fig. 7.5 Field of view in a telescope fitted with a filar micrometer

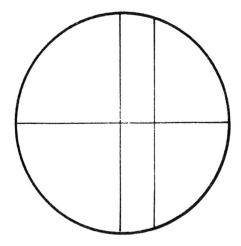

height of the sun (*A*) over the feature at the time of observation is now computed from the following formula:

sin *A* = (sin *B*)(sin *B*″) + (cos *B*)(cos *B*″) sin (*c* + *L*)

At the time of the shadow length observation the Earth and Sun subtend an angle at the Moon's center called the Auxiliary Angle and designated *F* in the following formula for calculating it:

cos *F* = (sin *B*′) (sin *B*″) + (cos *B*′)(cos *B*″) sin (*L*′ + *C*)

Finally, the height of the feature (*H*) may be calculated as a decimal of the radius of the Moon using the formula:

H = *D* (sin *A*) (cosec *F*) − ½ (*D*²) (cosec ²*F*) (cos ²*A*)

(where *D* is the shadow length as a decimal fraction of the Moon's radius). This may easily be converted to miles or kilometers if desired. Now that programmable pocket calculators with trigonometrical functions are easily available, the calculations necessary for determining the heights of lunar features from the raw data are no longer as fearsome as they once were. The determination of lunar heights may be involved in the following types of investigations.

i Intensive studies of localised areas. The amateur who intensively studies a localised area such as the crater Posidonius would do well to include shadow length measurements in the observing program. If the height of an outstanding feature is known in or near the area, then it is a simple matter to measure the heights of other objects in the vicinity by applying Mädler's formula:

H/*H*₀ = *S* sin*A*/(*S*₀ sin *A*₀)
where
H is the height to be determined,
S is the length of its shadow,
A is the angular altitude of the sun in the position of the feature, at the time of observation,
*H*₀,*S*₀ and *A*₀ are the corresponding values for the standard object.

ii Heights of ridges in maria. Most mare ridges are less than 500 feet high, therefore their shadows should be measured when they are very close to the terminator. Really accurate coordinates for the observed points will need considerable care. The measurement of marial ridge heights is a useful pursuit as there is only little in the way of published data.

iii Fault heights. Height studies of faults such as those in the craters Boscovitch and Bürg and the one near Cauchy would be valuable, as the only one on which there appears to be detailed height information is the Straight Wall.

iv Project 'Moonhole'. Some years ago the British Astronomical Association initiated a project, the object of which was the reverse of height measurement, namely the measurement of crater depths by shadow length methods.

Attention was concentrated on those with well-rounded interiors so that accurate estimates of depths could be made, noting what fraction of the crater diameter was taken up by the interior shadow. Knowing the altitude of the Sun, the depth of the crater could be readily calculated. This method was used for craters 15 miles or less in diameter and was designated 'Project Moonhole' but it was abandoned as being insufficiently accurate. However, crater *profiles* may be profitably studied. This is done by repeated observations of a given crater over several nights and study of the progress of its shadow. The height of the crater rim above the floor will vary as the edge of the shadow moves across the crater so enabling the crater profile to be determined. Of particular value would be determination of the profiles of the outer slopes of craters.

Further Reading

Books – General

Guide for Observers of the Moon. British Astronomical Association (BAA) (1979).
> This booklet contains detailed instructions for the study of small lunar crater profiles, transient lunar phenomena, the construction of a 'Moon blink' device, observation of occultations and lunar photography.

Observational Astronomy for Amateurs. 4th Edition. Sidgwick, J.B. Enslow Publishers, New Jersey, United States of America (1982).
> The 3rd Edition (Faber and Faber, London, England 1971) of this book is reprinted as a paperback by Dover Publications, Inc., New York, United States of America (1980).

Handbook of Practical Amateur Astronomy. Moore, P.A. (ed.). W.W. Norton and Co., Inc., United States of America (1964). Chapter 6.

The Amateur Astronomer's Handbook. Muirden, J. Thomas Y. Crowell Co., New York, United States of America (1974). Chapter 6.
> This book is also published in Great Britain entitled *Astronomy for Amateurs*.

Amateur Astronomy. Moore, P.A. W.W. Norton and Co., Inc., United States of America (1968). Chapters 6 and 7.

Papers and Articles (Selected Topics)

'Lunar Incognita: The Last Frontier?' Westfall, J.E. *The Strolling Astronomer*, **23** (7–8), 118–22 (1972).

'The Lunar South Polar Regions.' Whitaker, E.A. *JBAA*, **64**, 234–42. (1953/4).

'On Visual Estimates of the Depths of Small Craters.' Warner, B. *JBAA*, **73**, 224–8 (1962/3).

'A Request for Observation of Central Peaks.' Chapman, C.R. *The Strolling Astronomer*, **17** (7–8), 162–3 (1963).

'What is a Lunar Dome?' Ashbrook, J. *The Moon* (BAA Lunar Section Bulletin), **8**(2), 30–2 (1960).

'Lunar Domes.' Rackham, T. *The Moon* (BAA Lunar Section Bulletin), **7** (4), 71–2 (1959).

'Lunar Domes.' Moore, P.A. *Sky and Telescope*, **18** (1), 91–5 (1958/9).

'The Lunar Dome Survey: A Progress Report.' Jamieson, H.D. *The Strolling Astronomer*, **23** (11–12), 212–15 (1972).

'Fault Scarp in Sinus Roris: Observations and Comments.' Jamieson, H.D. *The Strolling Astronomer*, **17** (7–8), 167–9 (1963).

'The Bright Rays on the Moon.' Roth, G.D. *The Strolling Astronomer*, **6** (12), 171–6 (1952).

'The Copernicus Ray System.' Lenham, A.P. *JBAA*, **65** 347–8 (1954/5).

'Observation of a Volcanic Process on the Moon.' Kozyrev, N.A. *Sky and Telescope*, **18** 184–6 (1958/9).

'Report on Transient Phenomena.' Moore, P.A. *JBAA*, **77** 47–9 (1966/7).

'Transient Lunar Phenomena: A Review, 1967.' Moore, P.A. *JBAA*, **78**, 138–44 (1967/8).

'Lunar Transient Phenomena. The ALPO Program.' Ricker, C.L. *JBAA*, **78** 217–19 (1967/8).

'Extension of the Chronological Catalogue of Reported Lunar Events: October 1967–June 1971.' Moore, P.A. *JBAA*, **81** 365–90 (1970/1).

'Transient Lunar Phenomena – A New Approach.' Fitton, L.E. *JBAA*, **85** 511–27 (1974/5).

'Results from the ALPO–TLP Observing Program.' Cameron, W.S. *JBAA*, **89** 454–64 (1978/9).

'List of Dark Patches on the Moon.' Pickering, W.H. *JBAA*, **74** 197–9 (1963/4).

'Areas on the Moon Suspected of Variability.' Moore, P.A. *JBAA*, **75** 119–27 (1964/5).

'Lunar Notes.' (Drawings by Inez Beck and Notes of the Variable Dark Markings in Eratosthenes.) *The Strolling Astronomer*, **22** (9–10), 175–77 (1970).

'Periodic Changes in the Brightness of Four Light Spots in the Lunar Crater Schickard.' Emley, E.F. *JBAA*, **48** 76–9 (1937/8).

'Variations Within Plato.' Wilkins, H.P. *JBAA*, **53** 190–2 (1942–3).

'Three Riddles of Plato.' Carle, J.T. *Sky and telescope*, **14** 221–3 (1954/5).

'A Plato Illusion.' Ashbrook, J. *Sky and Telescope*, **19** (2), 92 (1959/60).

'Illusions That Trap Lunar Observers.' Copeland, L.S. *Sky and Telescope*, **15** (6), 248–51 (1955/6).

'Variations in the Lunar Formation Aristarchus.' Wilkins, H.P. *JBAA*, **56** 12–14 (1945/6).

'The Herodotus Puzzle .' Hartmann, W.K. *Sky and Telescope*, **16** (9), 451 (1956/7).

Strange World of the Moon. Firsoff, V.A. Hutchinson and Co., London, England (1959). Chapters 8 and 9.

The Old Moon and the New. Firsoff, V.A. Sidgwick and Jackson, London, England (1969). Chapter 13.

'Lunar Banded Craters.' Abineri, K.W. and Lenham, A.P. *JBAA*, **65** 160–6 (1955).

'Banded Craters.' Robinson, L.J. *JBAA*, **73** 33–8 (1962/3).

'A Survey of Dark Lunar Radial Bands.' Leatherbarrow, W.J. *JBAA*, **77** 33–8 (1966/7).

'The Bands of Aristarchus.' Barker, R. *JBAA*, **58** 99–101 (1947–8).

'Features of the Lunar Crater Aristarchus.' McIntosh, R.A. *JBAA*, **71** 380–8 (1961). (This paper contains a series of drawings of Aristarchus showing its appearance from sunrise to sunset.)

'Lunar Changes.' Pickering, W.H. *7th Report of the Section for the Observation of the Moon*. BAA, London, England. (1916).

'Physical Changes on the Moon.' Barker, R. *JBAA*, **48** 347–53 (1937/8).

'A Lunar Physical Change.' Barker, R. *JBAA*, **59** 181–2 (1948/9).

'Some Lunar Oddities and Curiosities.' Warner, B. *The Moon*. (BAA Lunar Section Bulletin), **7** (4), 87–8 (1959).

'Does Anything Ever Happen on the Moon?' Haas, W.H. *Journal of the Royal Astronomical Society of Canada*, **36**: (6) 237–72; (7) 317–28; (8) 361–76; (9) 397–408 (1942).

'Colour on the Moon.' Moore, P.A. *JBAA*, **74** 225–9 (1964).

'Color on the Moon.' Firsoff, V.A. *Sky and Telescope*, **17** 329–31 (1958).

'Lunar Colors.' Avigliano, D.P. *The Strolling Astronomer*, **8** 50–5 (1951).

'Color Changes on the Moon.' Haas, W.H. *Popular Astronomy*, **45** 337 (1937).

'Colour Filter Observation.' Heath, A.W. *The Moon* (BAA Lunar Section Bulletin), **9** (4), 73 (1961).

'Colour Filter Report.' Moore, P.A. *JBAA*, **74** 139–47 (1963/4).

Strange World of the Moon. Firsoff, V.A. Hutchinson and Co., London, England (1959). Chapter 10.

The Old Moon and the New. Firsoff, V.A. Sidgwick and Jackson, London, England (1969). Chapter 4.

'Next Month's Lunar Eclipse.' Observer's Page. Abileah, R. *Sky and Telescope*, **41** 55–7 (1971).
  This article contains useful instructions for observing and recording lunar eclipses.

'An Experiment on Penumbral Lunar Eclipses.' Fujinami, S. and Yamasaki, Y. *Sky and Telescope*, **18** 620–1 (1959).

'Observation of Occultations.' Taylor, G.E. In *Handbook of Practical Amateur Astronomy*. P.A. Moore (ed.). W.W. Norton and Co., Inc., United States of America (1964). Chapter 7.

'Predicting and Observing Lunar Occultations.' Sadler, F.M. *Sky and Telescope*, **19** 84–6 (1959).

'Lunar Photography for Beginners.' Hatfield, H. *JBAA*, **76** 90–102 (1965/6).

'Some Notes on Lunar Photography with a 4¼-inch Refractor.' Hill, J.J. *JBAA*, **55** 77 (1944/5).

APPENDIX 1   Numerical data pertaining to the Moon

Diameter:	2160 miles
Surface area:	1.46×10^7 square miles
Mass:	0.01227 Earth
Volume:	0.0203 Earth ($5,230 \times 10^9$ cubic miles)
Density:	0.606 Earth (3.34 water)
Surface gravity:	0.166 Earth
Escape velocity:	1.5 miles/second
Mean eccentricity of orbit:	0.0549
Distance from Earth:	Maximum (Apogee) 252 948 miles
	Minimum (Perigee) 221 593 miles
Apparent diameter:	Perigee (maximum) 33'31"
	Apogee (minimum) 29'22"
	Mean 31'5"

Orbital velocity (mean):   0.63 miles/second

Sidereal period (mean):   27 days 7 hours 43 minutes 11.47 seconds (27.3217 days)

Synodic period: 29 days 12 hours 44 minutes 2.78 seconds (29.5306 days)

Inclination of orbit (mean) relative to Ecliptic: 5°8'43"

Inclination of axis to orbit: 1°30"

Revolution of nodes: 18 years 10 days 8 hours 30 minutes

Saros: 18 years 10 days 7 hours 42 minutes

Motion of terminator in longitude per day (mean):   11.49° = 216.4716 miles at the Equator

One degree of Longitude at the Equator = 18.84 miles

Area of Moon always seen from Earth: 0.589 of whole surface

Area of Moon never seen: 0.411 of whole surface

Area of libratory regions: 0.178 of whole surface

Radius of polar regions: 24 miles

Atmospheric density (maximum possible): 1×10^{-4} Earth's

Distance from limb to limb: 3391.2 miles

Surface temperature: Maximum +118°C

Minimum −153°C

Albedo: 0.073

Color index: $+1.18^{m}$

Apparent magnitude: Full Moon -12.55^{m}
Half Moon -10.20^{m}

APPENDIX 2   Interesting facts

Largest walled formation	Bailly, diameter 183 miles
Deepest crater	Newton, depth from peak to floor: 29 000 feet
Highest central hill	In Moretus, height 7700 feet
Largest central hill	In Alpetragius
Largest mare	Mare Imbrium
Highest mountain	Beta in the Leibnitz Range, height 36 000 feet
Longest cleft	Sirsalis, length 300 miles
Longest white rays	Tychonian system
Steepest precipices	Crest of Copernicus
Brightest area	Interior of Aristarchus
Darkest area	Floor of Grimaldi
Roughest area	South-west
Smoothest area	North-east

APPENDIX 3   Some dates

First telescopic observation of the Moon (Galileo)	1610
First map of the Moon with names (Langrenus)	1645
Hevelius's map	1647
Riccioli's map	1651
Schröter's *Selenotopographische Fragmente*	1791,1802
Lohrmann's *Sections*	1824
Mädler's map	1837
First photograph of the Moon (Draper)	1840, March 23
Neison's map	1876
Schmidt's map	1878
Elger's map	1895
Pickering's photographic atlas	1904
Goodacre's map	1910
IAU map	1935
Wilkins's map	1946
Kuiper's photographic atlas	1960
Consolidated Lunar Atlas (photographic)	1967
Lunar probes	
First probe to reach the Moon (Luna 2, USSR)	1959, September
First photographs of Moon's averted hemisphere (Luna 3, USSR)	1959, October
First close-up photographs of surface (Ranger 7)	1964, July 31
First soft landing of probe (Luna 9, USSR)	1966, February 3
First men on the Moon (Apollo 11)	1969, July 20
Apollo 12	1969, November 19
Apollo 14	1971, February 5

Apollo 15	1971, July 30
Apollo 16	1972, April 21
Apollo 17	1972, December 11

APPENDIX 4   Books about the Moon

A selection of books about the Moon will be found at the end of Chapter 1. These are of a general nature and will provide the reader with plenty of 'background knowledge'. Other books of this type will be found listed at the ends of Chapters 2 and 4.

Of particular interest to the practical lunar observer will be the books that contain detailed descriptions of the Moon's surface features accompanied by sectional maps. A list of these is given here. Unfortunately, these works are all out of print and those of Elger, Neison and Goodacre are now quite rare but copies should be found in most astronomical libraries.

The Moon and the condition and configuration of its surface. Neison, E. Longmans, Green & Co. (London), 1876.
> The text is based mostly on Mädler's *The Moon* (1837) with all other available data integrated with it. The accompanying map, diameter 24 inches, in 22 sections, was based on Beer and Mädler's *Mappa Selenographica* with detail added from Neison's own observations.

The Moon. A Full Description and Map of its Principal Physical Features. Elger, T.G. George Philip & Son (London), 1895.
> This was the first English reference work not based on Mädler. It includes Elger's 18-inch map of the Moon in four quadrants. The map was later revised by H.P. Wilkins.

The Moon with a Description of its Surface Features. Goodacre, W. Published privately by the author, Bournemouth, England, 1931.
> The book is illustrated by a reduced version of Goodacre's original map which was on a scale of 1 : 18 000 000 published in London in 1910 in 25 sections. About 1400 positions measured by Saunder were employed in constructing the map which was based on photographs and the first to use the rectangular grid. Its positional accuracy was far superior to previous maps although its artistic quality was greatly inferior. Goodacre's map of the Moon is also reproduced in Hutchinson's *Splendour of the Heavens*, written by several authors and produced in 1923. Goodacre wrote the chapter on the Moon.

The Moon. A Complete Description of the Surface of the Moon, with the 300-inch Wilkins Lunar Map. Wilkins, H.P. and Moore, P.A. Faber & Faber (London), 1955.

Contains a reduced version of Wilkins's 300-inch map of the Moon in 25 sections and is illustrated with a unique set of drawings of selected lunar features by the authors from their observations with the great 33-inch Meudon refractor. Also included are charts of the libratory regions, the polar regions, a special chart of the walled plain Ptolemaus and a map of the Moon's averted hemisphere.

The Moon: Considered as a Planet, a World and a Satellite. Nasmyth, J. and Carpenter, J. John Murray (London), 1874.

Although this book does not contain a map of the Moon it is worthy of inclusion in this list. It ran into several editions, is mainly concerned with a defence of the Volcanic Hypothesis of the formation of lunar craters and is well worth studying. The photographic illustrations of Nasmyth's beautiful plaster models of lunar surface formations are noteworthy.

Celestial objects for common telescopes. Webb, T.W. Longman's, Green & Co. (London), 1859.

Contains a description of the Moon's surface based on Mädler's text, the first in English, accompanied by an outline map 11½ inches in diameter, based on the *Mappa Selenographica* of Beer and Mädler. Webb's book was reprinted by Dover Publications, New York, USA in 1962 as two paperback volumes with photographs.

APPENDIX 5   Maps and atlases of the Moon

(1) Non-photographic maps

Neison's Map.

This is 24 inches in diameter and appears in 22 sections. It was based on the *Mappa Selenographica* of Beer and Mädler with details added from Neison's own observations. The map is incorporated in Neison's book *The Moon* which was published in 1876 (see Appendix 4) but is now out of print.

Elger's Map

This originally appeared as a chart 18 inches in diameter in four quadrants in Elger's book *The Moon* which appeared in 1895 (See Appendix 4). It was revised by H.P. Wilkins and reissued by George Philip and Son Ltd (London) in 1959. It is also obtainable from the Sky Publishing Corporation, Cambridge, Mass., USA. The overall dimensions of the chart itself are 21 inches by 20 inches. Elger's revised map is reproduced on a scale of 13½ inches to the Moon's diameter on the *Solar System Giant Colorprint Map*, publication no. 9572 of the American Map Co. Inc., New York, NY, USA.

Goodacre's Map.

This was originally published in 1910 as a single sheet with a lunar diameter of 77 inches. Subsequently it appeared in a reduced version in 25 sections in Goodacre's book *The Moon*. Roughly 1400 positions measured by Saunder were used in constructing the map which was based on photographs and the first to use the rectangular grid. Its positional accuracy was far superior to previous maps although its artistic quality was greatly inferior. Goodacre's map is also reproduced in Hutchinson's *Splendour of the Heavens*, written by several authors and produced in 1923. Goodacre wrote the chapter on the Moon.

Mondkarte. Lohrmann, W.G. Johann Ambrosius Barth, Verlag (Leipzig), 1963.

A re-issue of Lohrmann's map published in 25 sections in 1878 by J.F. Julius Schmidt on a scale of 38.4 inches to the Moon's diameter. It is accompanied by a descriptive text in German and this with the 25 map sections are enclosed in a thick cardboard case. Lohrmann's map was the first detailed chart of the Moon prepared on a sound scientific basis.

Skeleton Map of the Moon. Rükl, A. Central Office of Geodesy and Cartography, Washington, DC.

This map is on a scale of 1 : 6 000 000 (diameter of lunar disc: 23 inches). In my opinion this is the best and most faithful of the smaller recent lunar maps. It is included with Kopal's *Photographic Atlas of the Moon* (see under (4)) in a pocket inside the back cover.

Map of the Moon. Moore, P.A.

Diameter of the lunar disc: 24 inches.

Sky and Telescope Lunar Map. Sky Publishing Corporation, Cambridge, Mass., USA.

A small map (diameter 10½ inches) but quite useful. Based on a drawing by Karel Andel published in 1926 entitled *Mappa Selenographica* with a superimposed grid of selenographic coordinates. There are 326 named formations.

NASA Map of the Moon. Issued in *Sky and Telescope*, November, 1970. A map of the near and far side of the Moon on Mercator's projection. There are two special charts of the polar regions. The scale is 1 : 10 000 000 At 34 degrees north and south latitudes and the map sheet itself measures 36¼ inches by 11½ inches.

(2) Photographic maps

Rand–MacNalley Official Map of the Moon. Freile, L. (ed.). Rand–MacNalley & Co., USA.

A very large chart (scale 1 : 23 000 000), 48 inches by 53 inches with a photographic representation of the Moon 47 inches in diameter. The lunar disc is oriented as seen with the naked eye with north at the top and south at the bottom. Fully labelled.

The Moon. Hallwag, Bern, Switzerland, 1969.

Another large chart (scale 1 : 5 000 000) 33 inches by 33 inches with a lunar diameter of 27 inches. Oriented as seen with the naked eye, north and south at top and bottom respectively. Fully labelled. There is a similar relief chart of the Moon's averted hemisphere on the reverse side. The accompanying booklet is in German and English.

(3) Non-photographic atlases

Moon Maps. Wilkins, H.P. Faber & Faber (London), 1960.

A revised and enlarged version of the author's map which appeared in *The Moon* by Wilkins and Moore. It is on a scale of 55.4 miles to the inch and the 25 sections are bound in a ring binder for use at the telescope. A special chart of the formation Ptolemaus is included and a sketch map – now obsolete – of the averted hemisphere. It is a pity that neither the positional accuracy nor artistic quality is commensurate with the amount of detail shown.

Moon Atlas. Firsoff, V.A. Hutchinson & Co. (London), 1961.

This incorporates a chart of the Moon drawn by the author, diameter 19 inches, in four quadrants, each with a gazeteer plus a relief map of the Moon, 22 inches in diameter and a 22-inch selenogical map. Includes photographs of lunar phases and the author's spherical projection photographs of the Moon's limb regions in which the foreshortening of the perspective is eliminated.

The Times Atlas of the Moon. Lewis, H.A.G. (ed.). Times Newspapers (London), 1969.
This work contains 109 relief maps of the Moon's surface on a scale of 20 miles to the inch. Unlike the previously mentioned atlases in which the Moon is represented as seen inverted in an astronomical telescope, the maps here are oriented with north and south at the top and bottom respectively. The maps have considerable aesthetic beauty but the coloring is pale. Surface relief is indicated by shading and the maps are drawn on a projection that eliminates foreshortening of surface features due to the Moon's curvature.

(4) Photographic atlases

Photographic Lunar Atlas. Pickering, W.H. Annals of Harvard College Observatory, 1903.
Shows each area of the Moon under five different illumination angles.

Photographic Lunar Atlas. Kuiper, G.P. University of Chicago Press, 1960. A boxed set of large (17 by 21 inches) photographs of the Moon's surface. The effect is somewhat heterogeneous as the atlas is compiled from a variety of sources (Mt Wilson, Lick, Pic du Midi, MacDonald and Yerkes Observatory photographs).

Orthographic Atlas of the Moon. Supplement no. 1 to the Photographic Lunar Atlas of G.P. Kuiper. University of Arizona Press, 1960.
Compiled by D.W.G. Arthur and E.W. Whitaker. Edited by G.P. Kuiper.
Sixty photographic plates with orthographic grid superimposed and orthographic coordinates. Hardcover, 18 inches by 23 inches.

Consolidated Lunar Atlas. Lunar and Planetary Laboratory, University of Arizona, 1967.
This is another boxed set of 226 large glossy photographs, each 11 inches by 14¼ inches, of the Moon's surface under different illumination conditions. The photographs were taken with the 61-inch NASA telescope of the Catalina Observatory and the 61-inch Astrometric telescope of the US Naval Observatory. The detail recorded is equivalent to what can be seen with an eight-inch telecope under good seeing conditions. There is an accompanying explanatory text. In my opinion these photographs are among the finest obtained with Earth-based telescopes.

Amateur Astronomer's Photographic Lunar Atlas. Hatfield, H. Lutterworth Press (London), 1968.
The photographs of the Moon's surface were all taken by the author with a telescope and camera of his own construction. The scale of the photographs corresponds to a lunar diameter of 25 inches. Identification of the surface formations in the photographs is made simple by the accompanying 16 labelled sectional maps. There is a total of 111 separate photographs and the lunar surface is shown under different illumination angles.

Photographic Atlas of the Moon. Kopal, Z. Academic Press Inc., 1965.
There are over 200 large photographs in this book, including a set of photographs of the Moon's phases. Rükl's *Skeleton Map of the Moon* is included in a pocket inside the back cover.

Pictorial Guide to the Moon. Alter, D. Thomas Y. Crowell Co. (New York), 1963.
110 Photographs, including lunar phases, 25 drawings and location key to lunar features. With text.

Lunar Atlas. Alter, D. Dover Publications Inc. (New York), 1964.
154 photographic plates, including lunar phases, and text.

Atlas of the Moon. De Callatay, V. MacMillan & Co. Ltd (London), 1964.
>   The 22 photographic plates of the Moon's surface in this book are accompanied by keys to features and explanatory text. There are many other photographic illustrations including lunar phases.

Photographic Lunar Atlas – No. 95 of Contributions from the Institute of Astrophysics and Kwasan Observatory, University of Kyoto, Japan.
>   This atlas consists of 85 plates prepared by S. Miyamoto and M. Matsui from photographs taken with the Kwasan refractor. The atlas is of handy reference size.

APPENDIX 6   Calculations of the Sun's selenographic colongitude

As explained in chapter 5, the Sun's selenographic colongitude (colongitude or colong. for short) is a precise measure of the position of the Moon's terminator at any phase and is independent of libration. It should be recorded for every observation and it is therefore important to know how to calculate it whenever an observation is made.

Colongitude is defined as, and is equal to, the eastern longitude of the sunrise terminator measured eastward from the Moon's central meridian all the way around the Moon. Hence, at mean libration, the colongitude at first quarter is 0°, at full Moon 90°, at last quarter 180° and at new Moon 270°. Due to libration in longitude these values may differ as much as 7.75° at the same apparent phases.

From this it follows that sunrise will occur at points east of the central meridian when the colongitude is equal to the eastern selenographic longitude of that point. Sunrise will occur at points west of the central meridian when the colongitude is 360° minus the western selenographic longitude at that point. For example, the crater Posidonius is located at 30.0° west. Sunrise begins at colongitude 360° − 30.0° = 330.0° while sunset occurs at colongitude 330.0° + 180° − 360° = 150.0°.

The colongitude at 0.00 hrs UT (Universal Time) is listed for every day of the year in the *Handbook of the British Astronomical Association* and increases at a nearly uniform rate of 0.50791° per hour.

Example of the calculation of colongitude

What is the colongitude at 10.50 p.m. EST (Eastern Standard Time, USA) on April 11, 1984?

First, convert 10.50 p.m. EST (22.50 hrs) to UT (Universal Time). EST is 5 hours behind UT, therefore 10.50 p.m. EST on April 11 is 22.50 + 5.0 = 27.50 hrs = 03.50 hrs on April 12 (notice that conversion to UT may change the date).

On consulting the BAA *Handbook* we see that the colongitude at 0.00 hrs UT on April 12 is 40.7°. We want the colongitude for a time that is 3 hours and 50 minutes later than this. From the adjoining table we see that colongitude increases by 1.94°

during this time. We simply add this to the value for 0.00 hrs and this gives us the colongitude we want:

Colongitude at 3.50 UT on April 12, 1984
= Colong. at 0.00 UT + 1.94°
= 40.7° + 1.94°
= 42.64°

Colongitude intervals

hr	0	5	10	15	20	25	30	35	40	45	50	55
1	.51	.55	.59	.64	.68	.72	.76	.81	.85	.89	.93	.98
2	1.02	.06	.10	.15	.19	.23	.27	.32	.36	.40	.44	.49
3	1.52	.56	.60	.65	.69	.73	.77	.82	.86	.90	.94	.99
4	2.03	.07	.11	.15	.20	.24	.28	.33	.37	.41	.45	.50
5	2.54	.58	.62	.67	.71	.75	.79	.84	.88	.92	.96	3.01
6	3.05	.09	.13	.18	.22	.26	.30	.35	.39	.43	.47	.52
7	3.56	.60	.64	.69	.73	.77	.81	.86	.90	.94	.98	4.03
8	4.06	.10	.14	.19	.23	.27	.31	.36	.40	.44	.48	/53
9	4.57	.61	.65	.70	.74	.78	.82	.87	.91	.95	.99	5.04
10	5.08	.12	.16	.21	.25	.29	.33	.38	.42	.46	.50	.55
11	5.59	.63	.67	.72	.76	.80	.84	.89	.93	.97	/6.01	6.06
12	6.09	.13	.17	.22	.26	.30	.34	.39	.43	.47	.51	.56
13	6.60	.64	.68	.73	.77	.81	.85	.90	.94	.98	/7.02	7.07
14	7.11	.15	.18	.24	.28	.32	.36	.41	.45	.49	.53	.58
15	7.62	.66	.70	.75	.79	.83	.87	.92	.96	/8.00	.04	8.09
16	8.13	.17	.21	.26	.30	.34	.38	.43	.47	.51	.55	.60
17	8.65	.69	.73	.78	.82	.86	.90	.95	.99	/9.03	.07	9.12
18	9.14	.18	.22	.27	.31	.35	.39	.44	.48	.52	.56	.61
19	9.65	.69	.73	.78	.82	.86	.90	.95	.99	/10.03	.07	.12
20	10.16	.20	.24	.29	.33	.37	.41	.46	.50	.54	.58	.63
21	10.67	.71	.75	.80	.84	.88	.92	.97	/11.01	.05	.09	.14
22	11.17	.21	.25	.30	.34	.38	.42	.47	.51	.55	.59	.64
23	11.68	.72	.76	.81	.85	.89	.93	/12.02	.06	.10	.15	.19

APPENDIX 7 Addresses of astronomical organisations

The American Lunar Society,
P.O. Box 209,
East Pittsburgh,
Pennsylvania, USA 15112

The Association of Lunar and Planetary Observers,
P.O. Box 16131,
San Francisco,
California,
USA 94116

The Royal Astronomical Society of Canada,
136, Dupont Street,
Toronto, Ontario,
Canada. M5R 1V2

The British Astronomical Association,
Burlington House,
Piccadilly,
London W1V 9AG,
England

APPENDIX 8   Published drawings and charts of the best known lunar formations

This list does not attempt to be comprehensive. As far as possible, drawings and charts from publications that are fairly easily accessible to most readers have been chosen. Out of print books and obscure literature have been avoided except in the case of drawings of outstanding merit or where there appears to be no drawing of a given formation in more accessible literature.

The list consists of drawings made mostly with the aid of moderate telescopes as well as several with large observatory instruments.

Formations on or very close (i.e., within one or two degrees of longitude) to the lunar limb are not included.

Abbreviations

JBAA: Journal of the British Astronomical Association
MBAA: Memoirs of the British Astronomical Association
BLS,BAA: Bulletin of the Lunar Section of the British Astronomical Association
W & M: Wilkins, H.P. and Moore, P.A. *The Moon* (Faber and Faber, London. 1961, 2nd edition)
Agrippa.   Porthouse, W. 8¼ in reflector. May 2, 1916. 8.15–9.45 p.m. *MBAA*, **20** (3), 7th Lunar Report 1916, plate V, fig. 15.
Alpetragius.   Wilkins, H.P. (telescope and date not stated). *The Moon (BLS,BAA)* **4** (2), 32 (1955).
Alphonsus.   Both, E.E. (Chart based on photographs and observations with 8 in refractor). *Journal of the International Lunar Society* **1** (6), 141 (1960).
Alpine Valley.   Elger, T.G. 8½ in reflector. January 23, 1885. *The Moon (BLS, BAA)*, **8** (2); front cover (1960).
Archimedes.   Hallowes, G.P.B. 12½ in reflector. January 14, 1924. *JBAA*, **24**, 132–3 (1923/4) (two other drawings are included).
Ariadaeus Cleft.   Fielder, G. 24 in refractor. December 12 1956. *The Moon (BLS,BAA)*, **5** (4), 72 (1957).
Aristarchus.   McIntosh, R.H. 14 in reflector. (Several drawings, sunrise to sunset.) *JBAA*, **71** (8), 381,383,387 (1961).
Aristarchus, Herodotus and Schröter's Valley.   Wilkins, H.P. 14¼ in reflector. March 30, 1950. *A Guide to the Moon*. Moore, P.A. W.W. Norton & Co (1953), plate facing p. 81. Ball, L.F. (telescope and date not stated). *Survey of the Moon*. Moore, P.A. Eyre & Spottiswoode (1963), plate facing p. 176.

Arzachel. Goodacre, W. (Drawing based on 100 in telescope photograph.) *MBAA*, **23** (4), 8th Lunar Report, p. 101.

Atlas. Molesworth, P.B. 12½ in reflector. August 10, 1892. *MBAA*, **7** (3), 4th Lunar Report 1899, plate 1, fig 3.

Bailly. Lenham, A.P. and Abineri, K.W. *JBAA*, **66**, plate facing p. 26 (1955/6).

Bartlett (Mädler's Square). Moore, P.A. 33 in refractor. April 22, 1953. *W & M*, p. 226.

Bonpland (and Parry). Herring, A.K. 12½ in reflector. June 26, 1958. 04.50 UT. *Sky and Telescope*, **19**, 32–35 (1959/60).

Bullialdus. Ball, L.F. (telescope and date not stated). *A Guide to the Moon*. Moore, P.A. W.W. Norton & Co., 1953, plate facing p. 80.

Capuanus. Wessling, R.J. 12½ in reflector. October 29, 1971. 0.47–2.00 UT. *The Strolling Astronomer*, **23** (11–12), front cover (1972).

Cassini. Herring, A.K. 12½ in reflector. July 24, 1958. 04.30 UT. *Sky and Telescope*, **18**, 515 (1959).

Catharina. See under Theophilus, Cyrillus and Catharina.

Clavius. Wilkins, H.P. (chart). *MBAA*, **36** (3), 11th Lunar Report 1950, p. 32.

Cleomedes. Wilkins, H.P. 8½ in reflector. March 17, 1945. *JBAA*, **55**, 85 (1944/5).

Cobra Head. Moore, P.A. 33 in refractor. April 7, 1952. *W & M*, p. 263.

Copernicus. Elger, T.G. 8½ in reflector. March 11, 1889. *Sky and Telescope*, **25**, 248 (1963).

Crisium, Mare. Hill, H. (chart of south part). *MBAA*, **36** (3), 11th Lunar Report 1950, p. 6.

Cyrillus. See under Theophilus, Cyrillus and Catharina.

Darwin. Wilkins, H.P. 15¼ in reflector. February 5, 1955. *JBAA*, **65**, plate 11 facing p. 189 (1954/5).

Diophantus and Delisle. Cooke, S.R.B. 12½ in reflector. August 9, 1935. *MBAA*, **36** (1), 10th Lunar Report 1947, plate IV.

Endymion. Goodacre, W. (chart). *MBAA*, **32** (2), 9th Lunar Report 1936, p. 8.

Eratosthenes. Collinson, E.H. (several drawings). 3 in refractor. *JBAA*, **34**, 306–8 (1923/4).

Fracastorius. Elger, T.G. 8½ in reflector. September 20, 1883. *The Moon (BLS,BAA)* **5** (1), 13 (1956).

Gassendi. Abineri, K.W. (Two drawings, morning and evening illumination.) 8 in reflector. *The Moon (BLS,BAA)*, **3** (2), 37 (1954).

Goldschmidt. Goodacre, W. 10 in refractor. May 4, 1933. 18hr 30min. *JBAA*, **43**, 324 (1932/3).

Grimaldi. Abineri, K.W. 8 in reflector. December 31, 1960. 11.30–13.00 GMT. *JBAA*, **75**, 314,315 (1964/5).

Guericke. Wilkins, H.P. 33 in refractor. April 3, 1952. 22.00–22.15 UT. *W & M*, p. 217.

Hainzel. Goodacre, W. (Drawing based on 100 in telescope photograph.) *MBAA*, **32** (2), 9th Lunar Report 1936, p. 10.

Hercules. Hallowes, G.P.B. July 29, 1915. 12.00–13.30. *MBAA*, **20** (3), 7th Lunar Report 1916, plate VIII.

Herodotus. See Aristarchus, Herodotus and Schröter's Valley.

Herschel, J. Moore, P.A. 33 in refractor. April 24, 1953. *W & M*, p. 245.

Hesiodus. Herring, A.K. 12½ in reflector. May 24, 1961. 04.35 UT. *The Moon (BLS,BAA)*, **10** (1), 13 (1961)

Hevel. Cooke, S.R.B. 10½ in reflector. September 10, 1935. *MBAA*, **32** (2), 9th Lunar Report 1936, p. 12.

Humorum, Mare. Lenham, A.P. (Chart, observations with 3 in refractor.) *JBAA*, **60**, 167–8 (1949/50).

Hyginus Cleft. Whitaker, E.A. (Drawing based on Pic-du-Midi photograph.) March 21, 1945. *The Moon (BLS, BAA)*, **4** (4), 77 (1956).

Iridum, Sinus. Wilkins, H.P. (Telescope and aperture not given.) *W & M*, plate facing p.246.

Janssen. Cooke, S.R.B. 12 in reflector. 1927. *JBAA*, **56**, plate between pp. 142,143 (1945/46).

Julius Caesar. Elger, T.G. 8½ in reflector. 1891. *The Moon (BLS,BAA)*, **9** (3), front cover (1961).

Kepler. Herring, A.K. 12½ in reflector. January 26, 1964. 02.10 UT. *Sky and Telescope*, **30**, 50 (1965).

Letronne. Elger, T.G. 8½ in reflector. May 21, 1888. 20.30–21.30 UT. *Sky and Telescope*, **25** 248 (1963).

Linné. Corder and Goodacre, 1900; Fauth, 1906; Wilkins and Moore, 1953. (Three drawings by different observers.) *JBAA*, **64**, 87 (1953/4).

Mädler's Square. See Bartlett.

Marius. Herring, A.K. 12½ in reflector. August 20, 1964. 06.40 UT. *Sky and Telescope*, **29**, 59 (1965).

Messala. Green, S.M. *MBAA*, **36** (1), 10th Lunar Report 1947, plate I.

Messier and Pickering. Daniels, M. 8 in reflector. Undated. *Astronomy*, **12** (11), 83 (1984).

Moretus. Abineri, K.W. and Lenham, A.P. (Chart.) *JBAA*, **65**, 82 (1954/5).

Nasmyth. See Phocylides.

Palitzsch. Moore, P.A. 25 in refractor. October 4, 1952. *W & M*, p. 172.

Parry. See Bonpland.

Petavius. Wilkins, H.P. 30 in reflector. October 5, 1952. *W & M*, p. 174.

Phocylides (and Nasmyth). Abineri, K.W. 8 in reflector. October 28, 1955. 8.45–11.00 GMT. *The Moon(BLS,BAA)*, **4** (3), 61 (1956).

Pitatus. Herring, A.K. 12½ in reflector. January 1, 1958. 05.15 UT. *Sky and Telescope*, **17**, 637 (1957/8).

Plato. Wilkins, H.P. 33 in refractor. April 3, 1952. *W & M*, p. 234.

Posidonius. Herring, A.K. 12½ in reflector. September 13, 1957. 09.05 UT. *The Moon (BLS,BAA)*, **6** (2), 23 (front cover) (1958).

Ptolemaus. (Chart.) *W & M*, p. 61.

Rheita Valley. Abineri, K.W. (Chart.) *JBAA*, **65**, 168 (1954/5).

Riccioli. Wilkins, H.P. 33 in refractor. April 8, 1952. *W & M*, p. 272.

Schickard. Abineri, K.W. 8½ in reflector. *W & M*, plate facing p. 300.

Schröter's Valley. See Aristarchus, Herodotus and Schröter's Valley.

Sirsalis Cleft. (North.) Ford, A. 8½ in reflector. March 22, 1959. 22.00–24.00. *The Moon (BLS,BAA)*, **9** (2), 35 (1961).

South Polar Region. Whitaker, E.A. (Chart.) *W & M*, p. 320.

Stadius. Wilkins, H.P. 33 in refractor. April 22, 1953. *W & M*, p. 118.

Stöfler Wilkins, H.P. (Chart.) *JBAA*, **52**, 64 (1941/2).

Straight Wall. Wilkins, H.P. 26 in refractor. *W & M*. plate facing p. 146.

Theophilus. Bolton, S. (Composite drawing and chart.) *MBAA*, **13** (3), 6th Lunar Report, 1906, p. 82 and plate II.

Theophilus, Cyrillus and Catharina. Elger, T.G. 8½ in reflector. March 23, 1893. 20.00–21.30 UT. *Sky and Telescope*, 25,249 (1963).

Triesnecker Clefts. Cyrus, C.M. 10 in reflector. July 21, 1961. *The Strolling Astronomer*, **16** (5–6), front cover (1962).

Tycho. Elger, T.G. 8½ in reflector. 1888. *The Moon (BLS,BAA)*, **9** (4), 63 (front cover) (1961).

Vendelinus. Wilkins, H.P. 30 in reflector. December 3, 1952. 22.15–23.00 UT. *JBAA*, **63**, plate XI facing p. 116 (1953).

Walter. Elger, T.G. 8½ in reflector. March 9, 1889. 19.45–20.30 UT. *Sky and Telescope*, **25**, 248 (1963).

Wargentin. Wilkins, H.P. and Moore, P.A. 33 in refractor. April 7, 1952. *W & M*, p. 302.

Subject Index

Author and Name Index

Index of Named Lunar Surface Features

(Illustrations are indicated in bold face type)